Problem Solving in
Clinical Anatomy

For Churchill Livingstone

Commissioning Editor Timothy Horne
Copy Editor Isla MacLean
Project Controllers Kay Hunston, Debra Barrie
Design Direction Erik Bigland
Sales Promotion Duncan Jones

Problem Solving for Tutorials in
Clinical Anatomy

Peter H. Abrahams MBBS FRCS(Ed) FRCR
Clinical Anatomist, University of Cambridge
Examiner in Anatomy to The Royal College of Surgeons of Edinburgh
Former Examiner to The Royal College of Surgeons of England
Fellow of Girton College, Cambridge

Matthew J. Thatcher FRCS(Eng)
FRCS (Ed) DMRD DRCOG
Senior Medical Officer, Department of Health, London

Jonathan D. Spratt MA MB BChir
Senior House Officer in General Surgery, Bath Royal United Hospital
Former Demonstrator in Anatomy, University of Cambridge

SECOND EDITION

CHURCHILL LIVINGSTONE
EDINBURGH HONG KONG LONDON MADRID MELBOURNE
NEW YORK AND TOKYO 1995

CHURCHILL LIVINGSTONE
Medical Division of Pearson Professional Limited

Distributed in the United States of America by Churchill Livingstone Inc., 650 Avenue of the Americas, New York, N.Y. 10011, and by associated companies, branches and representatives throughout the world.

© Peter H. Abrahams, Matthew J. Thatcher and Jonathan D. Spratt 1995

All rights reserved. No part of this publication may be reproduced, stored in a retrieval system, or transmitted in any form or by any means, electronic, mechanical, photocopying, recording or otherwise, without either the prior permission of the publishers (Churchill Livingstone, Robert Stevenson House, 1–3 Baxter's Place, Leith Walk, Edinburgh EH1 3AF), or a licence permitting restricted copying in the United Kingdom issued by the Copyright Licensing Agency Ltd, 90 Tottenham Court Road, London, W1P 9HE

First published 1981 (Pitman Publishing Ltd)
This edition 1995

ISBN 0 443 05285 9

British Library Cataloguing in Publication Data
A catalogue record for this book is available from the British Library

Library of Congress Cataloging in Publication Data
A catalog record for this book is available from the Library of Congress

First published with the title *Pocket Examiner in Regional and Clinical Anatomy*.

The publisher's policy is to use paper manufactured from sustainable forests

Produced by Longman Singapore Publishers (Pte) Ltd
Printed in Singapore

Contents

Preface vii
Acknowledgements ix

QUESTIONS
Head and neck 1
Upper limb 16
Thorax 24
Abdomen 31
Pelvis and perineum 41
Lower limb 48
Back 55

ANSWERS
Head and neck 59
Upper limb 114
Thorax 138
Abdomen 165
Pelvis and perineum 212
Lower limb 238
Back 271

Biographical notes on eponyms 281

Preface

Examinations and assessments are necessary evils for students and teachers alike. None causes more anxiety than the oral, or viva, examination when, often for the first time, examiner and candidate meet face to face. Preparing for both oral and written examinations in anatomy is a long process, there being no short cuts to learning the facts.

However, having analysed both written and oral undergraduate and postgraduate examinations over some years, we have found a constant core of questions turning up in all medical schools and Royal College examinations. These tend to fall into one of two categories, in the same pattern as presented in this book: first, the information questions involving simple factual recall and, second, the applied questions, often involving a higher level of comprehension and sometimes a little digression from pure gross anatomy into the fields of embryology, radiology, surgery and medicine. This type of applied question tends to turn up more frequently in oral examinations, especially those for the FRCS and FRCR. It is extremely difficult to prepare for any oral examination except by question and answer practice, which we hope this book will stimulate, especially amongst small tutorial groups of students.

This book is intended to supplement and not to replace the standard texts on anatomy. However, many medical schools are directing their education towards clinical integration of the basic sciences and it has been found that these questions are ideal material for 'problem-solving' exercises. In some of the more traditional schools tutorials, supervisions and small group teaching are the norm. These questions have stimulated discussion between tutor and student as well as amongst the students themselves. As nothing is more soul-destroying, particularly in anatomy, than sitting alone trying to amass isolated facts, we hope that our efforts will make this task, and the frightening prospect of the anatomy oral or tutorial, a little less daunting!

P.H.A
M.J.T
J.D.S

Acknowledgements

We would like to thank all the students and colleagues who encouraged us to put these short questions and answers into a book format. Our special thanks go to Professor I. Glazer and Professor H. Gitlin, whose constructive criticism helped to consolidate our initial ideas during the participation of a WHO medical education project at Beer-Sheva, Israel. This contribution was made possible by the continuous support of Dr A. H. Taba, Director of the Eastern Mediterranean Regional Office of WHO, to whom we extend our gratitude. We would also like to thank Dr J. Walker for his biographical notes in the back of the volume, and for his help in proof reading.

To Miss Elisabeth Smale and Miss Anne Danagher go our special thanks for deciphering our awful handwriting and for typing the manuscript, and to Miss Tracey Ashley, Miss Pauline McLeod and Miss Tracey Marsh for their computing skills which have made the production of this new book such a pleasurable task.

Questions

HEAD AND NECK

A Bones of the skull

Information questions

1. Which skull bones form in membrane?

2. Which parts of the temporal bone develop in cartilage?

3. What is the metopic suture of the frontal bone?

4. What are the vertex, inion, glabella and crista galli?

5. The spine of the sphenoid has most important relations. What structures does this bony point protect?

6. Describe the foramina of the sphenoid bone. Which structures pass through them?

7. Where would you find the diploic veins?

8. What are emissary veins?

9. The conchae of the nose are formed by which bones?

10. Where is the foramen lacerum on a skull? Describe its relationship to the carotid artery.

11. What is the 'antrum' in the facial skeleton?

12. What class and type of joint is the basi-occipital/basi-sphenoid joint and when does it normally fuse?

Applied questions

13. How is the neonatal skull different from that of the adult?

14 What consequences during parturition do you think may result from the relatively small size of the fetal mastoid process?

15 A full sailor's beard overlies which bones?

16 A fracture in the region of the pterion may have serious consequences. What structure is at greatest risk in this injury?

17 What have the mastoid, ethmoid, sphenoid and maxilla in common?

18 Why might direct trauma and fracture of the hyoid bone prove dangerous?

19 Why might a fracture of the lower mandible cause numbness of the lower lip?

B Cranial fossae

Cranial meninges

Dural venous sinuses

Information questions

20 Where on a skull would you define the borders of the anterior and middle cranial fossae?

21 What structure lies between the clinoid processes?

22 What is the clivus, and which portion of the brain lies against it?

23 Which structures emerge from the skull through the jugular foramen?

24 Describe the relationship of the XIIth cranial nerve to the carotid arteries.

25 Where might you find the arachnoid granulations and lacunae within the skull?

26 What are the falces? Describe their functions.

27 Describe the contents of the cavernous sinus.

28 Which vessels drain into the cavernous sinus?

29 How would you find the superior sagittal sinus in a 6-month-old baby?

30 What vessels drain into the straight sinus?

31 Describe the walls of the superior sagittal sinus.

Applied questions

32 A patient presents with an increase in hat and glove size, and it is suggested that he may have a pituitary tumour. Which plain X-ray do you think might confirm this diagnosis?

33 A small depressed fracture of the vertex may cause thrombosis of the superior sagittal sinus. Why might this rapidly cause unconsciousness due to increased intracranial pressure?

34 On a slit Towne's X-ray of the skull, you should see the arcuate eminence of the temporal bone. What important structure causes this prominence?

35 A fracture through the pterygoid canal might damage which nerve fibres and with what consequences?

36 What is an important function of the tegmen tympani?

37 Why might a fractured cribriform plate of the ethmoid bone lead to a loss of smell (anosmia) and a runny nose (rhinorrhea)?

38 By which route might you aspirate a subdural haematoma from a small baby?

39 Why might pressing a boil of the cheek lead to a VIth nerve palsy?

40 Using X-rays, how can one view the major dural venous sinuses?

C Face

  Scalp

  Parotid region

Information questions

41 What have the platysma and the dartos muscle in common?

42 What is the sensory nerve supply to the skin over the angle of the jaw?

43 What is the sensory supply to the area of the upper lip covered by a moustache?

44 Describe the sphincter muscles of the face and their innervation.

45 What are the scalp's five layers?

46 Describe the arteries that anastomose in the scalp. From where do they originate?

47 Describe the secretomotor pathway to the parotid gland.

48 What arteries, veins and nerves lie within the parotid gland?

Applied questions

49 What are the effects on lacrimation of a damaged orbicularis oculi?

50 Describe the area affected by herpes zoster in the area supplied by the maxillary division of the trigeminal nerve.

51 Which are the most important muscles of facial expression?

52 How will the vestibule of the mouth be affected by a damaged VIIth cranial nerve?

53 Why might a lesion of the ophthalmic division of the trigeminal nerve be dangerous to the eye?

54 Where would a collection of blood or pus lie within the five layers of the scalp?

55 A young man with long flowing locks leans over his lathe and catches his hair in the machinery. Through which layer is he likely to be 'scalped'?

56 A generalised scalp infection will drain to which groups of lymph nodes?

57 How may an abscess on the scalp endanger the brain?

58 Why is mumps so painful?

59 Why might a cancer of the parotid gland present clinically as a facial paralysis?

60 If the auriculotemporal nerve is cut some distance beyond its origin, what will happen to the parotid gland?

61 Where would you look to find the opening of the parotid duct, and how might you palpate this duct on the face?

D Temporal and infratemporal regions

Temporomandibular joint

Information questions

62 In dissection, you often find the middle meningeal artery encircled by a small nerve. Which one?

63 What fibres does the nerve of the pterygoid canal (Vidian nerve) carry?

64 Describe the origin and course of the middle meningeal artery.

65 Which muscle protrudes and opens the jaw?

66 What is unusual about the structure of the temporomandibular joint?

67 Describe the movements of the temporomandibular joint.

68 What structures attach to, or hook round, the hamulus?

69 What structures attach to the pterygomandibular raphe?

Applied questions

70 To stop an extradural haemorrhage, which artery would you wish to tie and through which foramen does it enter the skull?

71 Why might an infection deep in the pterygoid region be of danger to the eye?

72 Why do babies who have a forceps delivery more commonly present with temporary facial palsies?

73 On clenching the jaw tight, what structure can you palpate lying on the contracted masseter?

74 Where does the dentist inject an anaesthetic if he wishes to numb your lower teeth?

75 Why is the pterygopalatine ganglion sometimes called the ganglion of 'hay fever'?

76 If you wished to visualise the articular tubercle, tympanic plate and external auditory meatus, what X-ray would you order?

77 Why might a dislocation of the temporomandibular joint occur during a boring anatomy lecture?

78 What muscles may be involved in trismus?

E Orbit

Eyeball

Information questions

79 Which bones form the orbit?

80 Which nerves pass through the superior and inferior orbital fissures, respectively?

81 Describe the fascial sheath of the eyeball (Tenon's capsule).

82 Which extraocular muscles originate from the tendinous ring?

83 What is the major refracting surface of the eye? Discuss its vascularity.

84 Name the six major extraocular muscles and their effects on the position of the eyeball.

85 Describe the muscles of the iris. What is their nerve supply?

86 Why is the optic nerve not a 'true' cranial nerve?

87 Where, on dissection, would you look for the ciliary ganglion?

88 Which fibres synapse in the ciliary ganglion?

89 What major differences are there between the long and short ciliary nerves?

Applied questions

90 What is the anatomical difference between a Meibomian cyst and a stye?

91 Explain two causes of a drooped upper eyelid (ptosis).

92 What is the anatomical explanation of swelling of the optic disc (papilloedema)?

93 How do you test accommodation, and which nerve fibres are involved?

94 Why is blockage of the central artery of the retina a disaster?

95 A patient comes into your clinic with ptosis, a dilated pupil and an eyeball that looks down and laterally. What might you diagnose?

96 If a patient came to you with progressive ptosis of the right eye but with normal examination of cranial nerves III, IV and VI, why might a chest X-ray be advisable?

97 Where is the blind spot in a normal eye?

98 If you use a drug to dilate the pupil for ophthalmoscopy, why might you cause an attack of raised intraocular pressure (acute glaucoma)?

99 A patient complains that he cannot turn his right eye to the right. What nerve lesion is the likely cause?

F Nose
Paranasal air sinuses

Palate

Tonsil

Pharynx

Information questions

100 The nasal septum is formed from which structures?

101 Describe the functions of the nasal cavity and also those attributed to the paranasal air sinuses.

102 What opens into the lateral wall of the nasal cavity?

103 Which arteries anastomose on the anteroinferior part of the nasal septum and from where do they originate?

104 What is the nerve supply to the soft and hard palates?

105 Describe the bed of the palatine tonsil and its arterial supply.

106 Describe the ring of lymphoid tissue in the upper pharynx (Waldeyer's ring).

107 Describe the fibres both passing through and synapsing in the pterygopalatine ganglion.

108 Why are the pharyngeal constrictor muscles often likened to a 'stack of flowerpots'?

109 The inferior constrictor has two parts. What is the difference in their origins?

110 Does the pharyngeal plexus supply all the muscles of the pharynx, palate and tongue?

111 Describe the motor contribution of the glossopharyngeal nerve and the one and only muscle it supplies directly.

112 Which muscles are involved in elevation of the hyoid during swallowing?

113 What are the valleculae?

Applied questions

114 Why are anterior epistaxes common and why are posterior epistaxes often more difficult to stop and may even require arterial ligation?

115 Why should a patient complaining about toothache be asked about recent upper respiratory tract infections and have his frontal sinuses examined?

116 During a posterior rhinoscopy examination, which structures would you hope to see?

117 What differences are there between the auditory tube (Eustachian, pharyngotympanic) of a child and that of an adult, and how does this help explain the frequency of middle ear disease in children?

118 To test which cranial nerves would you tickle the soft palate and make the patient 'gag'?

119 Why might a child with recurrent ear infections benefit from an adenoidectomy?

120 Your patient has tonsillitis. Which single palpable lymph node would you expect to be enlarged?

121 If you excise the epiglottis, what consequences do you think might result?

122 To what does the term 'hypopharynx' refer and, as it is not an uncommon site of cancerous growth, how could you examine it?

123 A pharyngeal pouch may occur, particularly in the elderly. What is its anatomical explanation?

124 Where is the most likely site for a fish bone to impact in the pharynx?

G Larynx

Thyroid

Anterior triangle of the neck

Information questions

125 Describe the embryological origins of the styloid process, stylomandibular ligament, hyoid bone and laryngeal cartilages.

126 Describe the laryngeal skeleton.

127 Describe the sphincter mechanisms of the larynx that protect the trachea from the entry of foreign bodies.

128 Which is the most vital muscle in the larynx?

129 Describe both the motor and the sensory nerve supply to the larynx.

130 Name the anatomical structure which is called the 'Adam's apple'.

131 What is the levator glandulae thyroideae?

132 Describe the blood supply of the thyroid gland.

133 Define the borders of the anterior triangle of the neck.

134 Where are the prevertebral and the deep cervical 'investing' fascia situated?

135 Describe the action of the infrahyoid and suprahyoid muscles on swallowing.

136 Why is the ansa cervicalis so called?

137 Name the branches of the external carotid artery and its two terminal arteries.

Applied questions

138 Why is the Adam's apple more prominent in the male, and what growth changes in the larynx accompany deepening of the male voice?

139 You are shown a lateral soft tissue X-ray of the neck and notice that some of the laryngeal skeleton appears as dense as bone and non-translucent. Can you suggest a reason for this occurrence?

140 At which vertebral levels do the following normally lie: hyoid, thyroid isthmus and cricoid cartilage?

141 When performing an indirect laryngoscopy, what do you think might be visible?

142 Discuss the effects on breathing and speech of an external laryngeal nerve palsy, and a bilateral paralysis of the recurrent laryngeal nerves.

143 In an anteroposterior tomogram X-ray of larynx, can you explain why the vocal cords are normally together?

144 What is the difference between tracheostomy and laryngotomy, and which important structures would you encounter performing these procedures?

145 Why does the inferior thyroid artery have such a tortuous course?

146 Why does the thyroid gland move on swallowing?

147 Why should a surgeon about to perform a thyroidectomy examine the vocal cords?

148 A patient sticks out his tongue and you notice that a swelling in the midline, above the hyoid, rises. What embryological abnormality do you suspect?

149 Why may a parathyroid gland be found within the thymus in a retrosternal position?

150 What are the contents and surface markings of the carotid sheath?

151 Discuss how you would easily distinguish the internal from the external carotid artery in the neck on a carotid arteriogram.

152 What important nerves might be damaged in an overenthusiastic carotid arteriogram puncture?

H Mouth

Tongue

Teeth

Mandible

Sublingual and submandibular salivary glands

Information questions

153 Describe the boundaries of the vestibule.

154 Which nerves are involved in general sensation (e.g. touch) from the tongue?

155 Describe the nerves involved in taste.

156 Describe the tongue muscles.

157 Describe the relationship of the tongue to the thyroid gland during development.

158 What is the chorda tympani, and what fibres does it carry?

159 What teeth would you expect to find in the mouth of a 4-month-old child, a 20-month-old child and a 12-year-old schoolboy?

160 Which nerves and glands are in direct contact with the mandible?

161 Describe the position of the two parts of the submandibular gland.

162 Describe the pathway of secretomotor parasympathetic impulses from the superior salivary nucleus to the submandibular gland.

163 What is the relationship between the lingual nerve and the submandibular duct?

Applied questions

164 Why might a young baby present with a small fleshy tumour in the midline at the back of the tongue? Describe its likely histology.

165 Why is the genioglossus muscle of importance in maintaining an airway in the unconscious patient?

166 Why might a patient with a carcinoma of the tongue complain of severe earache?

167 Describe why cancer of the posterior third of the tongue has a very bad reputation and reduced prognosis compared to that of the anterior two-thirds.

168 How might a patient present if she had a right hypoglossal nerve palsy?

169 Following extraction of the right lower wisdom tooth, a patient complained of a numb tongue and a metallic taste in the mouth. How can you explain this situation?

170 Why might a tumour of the middle ear cause abnormal sensations of taste?

171 In a group of 22-year-old students, how many teeth would you suppose the majority to have?

172 Why might a simple jaw fracture result in a numb lower lip?

173 Following removal of a submandibular gland for a cancerous growth, why might the patient have a 'drooped' lower lip?

174 Where do you think a radiologist would insert his cannula to do a sialogram of the submandibular gland?

I Posterior triangle of the neck

Occipital region

Information questions

175 Describe the floor of the posterior triangle.

176 Describe the borders and roof of the posterior triangle.

177 Where do lymph nodes lie on the side of the neck?

178 What sensory nerves emerge from the cervical plexus?

179 Where is the normal C1 dermatome situated?

Applied questions

180 Why may surgery in the inferior portion of the posterior triangle be very dangerous? What important structures might you encounter?

181 Why might a patient, following surgery for an abscess on the side of the neck, be unable to brush her hair?

182 Why might a known leprosy patient present with numbness over the lobule of the ear and a cord-like swelling along the posterior border of the sternocleidomastoid muscle?

183 Why is it relatively easy for infection in the occipital region to spread intracranially?

184 What important structure lies in the floor of the suboccipital triangle?

J Root of the neck
Sympathetic trunk
First rib

Information questions

185 What is the surface marking of the pleura in the root of the neck?

186 What are the branches of the subclavian artery in the neck?

187 Describe the course of the vertebral artery.

188 Can you explain the differences between the origins of right and left recurrent laryngeal nerves from their respective vagi?

189 What is the course of the thoracic duct from the cisterna chyli to the brachiocephalic vein?

190 Describe the formation and connections of the cervical sympathetic ganglia.

191 Describe the relationships of scalenus anterior, the phrenic nerve, transverse cervical artery and thoracic duct to each other.

192 Describe the relative positions of the lower trunk of the brachial plexus, subclavian vein and artery to the first rib.

193 How would you distinguish the first rib from any other rib?

Applied questions

194 Why might a brachiocephalic venous catheterisation be inadvisable on the left side?

195 Why might a gunshot wound in the region above the clavicle cause weakness in abduction of the shoulder?

196 Why is it dangerous to point the needle posteriorly when performing central venous catheterisation into the internal jugular, subclavian or brachiocephalic veins?

197 What is the stellate ganglion, and why might a cervical sympathectomy cause facial and ophthalmic signs (Horner's syndrome)?

198 What is a cervical rib, and how might it present in a patient?

K Ear

Information questions

199 Describe the sensory nerve supply to the pinna and the tympanic membrane.

200 Describe how sound is transmitted from the external ear to the fenestra vestibuli.

201 What is the important posterior relation of the mastoid antrum?

202 Describe the main features of the medial wall of the middle ear.

203 Describe the course of the VIIth cranial nerve within the petrous temporal bone.

204 What are the three small branches from the VIIth cranial nerve given off in the petrous temporal bone, and what is their functional importance?

205 Describe the membranous labyrinth.

206 What is the relationship between the three semicircular canals?

207 Describe the cochlea.

Applied questions

208 Describe the anatomy of the external ear from the exterior to the tympanic membrane and explain from where the wax comes.

209 What would you expect to see on the right tympanic membrane in a normal otoscopic examination?

210 How do you straighten the external auditory meatus in an adult, or in an infant, when trying to see the tympanic membrane?

211 When clearing their ears with cotton-wool sticks against medical advice, why do some people start coughing and a few people even vomit?

212 Why should Ramsay Hunt syndrome cause vesicles in the external auditory meatus?

213 A patient with abnormally sensitive hearing (hyperacusis), an unusual taste in the mouth, facial weakness, deafness and vertigo may well have a lesion in which region?

214 A young child presents in your clinic with a perforation of the eardrum and a large swelling in the mastoid region. Two days later he is brought into the clinic obviously very ill, drowsy and, from the mother's description, having just had convulsions. What do you fear has happened to his infection?

215 Can you explain why air hostesses give you sweets to eat at take-off and landing?

UPPER LIMB

A Axilla

Brachial plexus

Breast

Information questions

216 What spaces exist in the posterior wall of the axilla? What passes through them?

217 What is the axillary sheath?

218 Describe the base, apex, walls and contents of the axilla.

219 Describe the branches from the posterior and medial cords of the brachial plexus.

220 Describe the formation of the cords of the brachial plexus.

221 What is the distribution of dermatomes C5 and C6?

222 Describe the muscle bed of the female breast and how the normal breast maintains its rounded contour.

223 Describe the lymphatic drainage of the breast.

224 What is the 'boxer's muscle'?

225 Describe the blood supply to the female breast.

226 What do the terms 'prefixed' and 'postfixed' mean with respect to the brachial plexus?

Applied questions

227 What may be the neurological consequences of breaking the fall from a tree by grasping a branch?

228 Where can the axillary artery be surgically ligated without compromise to the blood supply of the upper limb?

229 What vessel is commonly injured in axillary stab wounds? What special dangers exist?

230 An invasive melanoma of the finger might spread to which lymph nodes of the axilla?

231 Why should a cervical rib cause loss of muscle bulk in the hand?

232 A small child was swung round very fast in a circle by his arms and afterwards was found to have sustained a traction injury to the upper trunk of the brachial plexus (Erb's palsy). In what position would the damaged arm be held and why?

233 A patient complained that, following a difficult radical mastectomy, her scapula stuck out like a wing when she pushed open a door. She also had difficulty in reaching the top shelves in her kitchen. Can you explain why?

234 You find that a patient still has sensation on the upper aspect of the shoulder following a gunshot wound in the region of the brachial plexus. Why might this be possible?

235 Why is an intramuscular injection posteriorly into the deltoid muscle potentially very dangerous?

236 What anatomical structure accounts for the fact that breast disease often causes dimpling and tethering of the skin?

237 Intradermal swelling and pitting (peau d'orange) is occasionally seen in cancers of the breast. What anatomical explanation can you suggest for this phenomenon?

238 What features might you expect to see on breast examination of a pregnant woman?

239 A patient has malignant change in the skin surrounding her nipple (Paget's disease). To which lymph nodes might metastases spread?

B Scapular region

Shoulder joint and girdle

Information questions

240 Describe the axillary artery and the arterial anastomoses around the scapula.

241 How would you distinguish between a right and a left clavicle?

242 Discuss the muscles involved in abduction of the arm to 180°.

243 What is the most important function of the coracoclavicular ligament?

244 What is the relationship of muscles in the intertubercular (bicipital) groove?

245 What factors contribute to the stability of the shoulder joint? Why should it dislocate so relatively easy?

246 Describe the rotator cuff muscles and their importance.

Applied questions

247 How may athletes dislocate their shoulder?

248 Discuss the joints of the clavicle. What are their common features?

249 What are the effects of a torn supraspinatus tendon?

250 What nerve might be damaged in a fracture of the neck of the humerus or a dislocation of the shoulder?

251 Where is the anatomical neck, as distinct from the surgical neck, of the humerus?

252 Why might a fractured clavicle appear with its medial third elevated and lateral two-thirds depressed and pulled anteriorly?

253 After a car accident your patient has a painful shoulder. Why might you ask especially for an axial X-ray of the shoulder in abduction?

254 What is meant by the term 'shoulder separation'?

C Arm

Elbow joint

Cubital fossa

Information questions

255 Describe the venous drainage of the upper limb.

256 Describe the courses and relationships, in the upper arm, of the median nerve and brachial artery. What happens to each as it leaves the cubital fossa?

257 What nerves are directly related to the humerus itself?

258 Describe the course of the cephalic vein in the arm.

259 What is the 'carrying angle'?

260 The elbow joint is really three different joints. What movements occur at each?

261 Describe the muscles involved in pronation and supination. Where is the axis of rotation?

262 Where can you normally palpate the ulnar nerve and the superficial branch of the radial nerve?

263 Describe the boundaries, roof, floor and contents of the cubital fossa.

264 Discuss the origins and distribution of the anterior and posterior interosseous nerves.

265 Describe the course of the radial nerve.

Applied questions

266 What is the surface marking of the brachial artery? Where may it be compressed to arrest haemorrhage in forearm/hand trauma?

267 Why is a damaged brachial artery a surgical emergency?

268 Describe the landmarks you would use in finding the brachial artery, either to measure blood pressure or to obtain arterial blood.

269 Why did a patient who had to use crutches for a week complain of difficulty in gripping things and a weakness at his wrists?

270 A famous boxer attempted an uppercut on an opponent and missed. Afterwards he noticed pain in the shoulder, a 'blacksmith's bulge' for his biceps and difficulty in turning a door handle. How do you explain this anatomically?

271 A supracondylar fracture of the humerus is a common injury in children who fall out of trees. Why might this be a serious injury and why may it possibly cause severe loss of muscle tissue distal to the fracture?

272 Why is a child's elbow relatively unstable?

273 Which bursae may become inflamed around the elbow joint?

274 What are the consequences, at the elbow, for golf and tennis enthusiasts?

275 A farmer falls over and cuts his ulnar nerve at the elbow. What functional effects will this have and how will his hand look some months later?

276 A young boy is dragged by the arm from an ice-cream shop and then complains of pain in the forearm. There is a small hollow 1 cm distal to the

lateral epicondyle. What is your diagnosis and why does this happen more often in children?

277 Why do screws have a right-hand thread?

278 Miners, and very occasionally nowadays students, may present with a swelling over the olecranon. What is it and how is it caused?

279 In a dislocated elbow joint, how would feeling the epicondyles and olecranon help in differential diagnosis from a supracondylar fracture?

280 On flexing his elbow, a patient complains of tingling in the little finger. An orthopaedic surgeon cures the problem by doing what procedure to which nerve?

281 A student has performed a clumsy venepuncture, pierced the 'grâce à Dieu' fascia and left the patient with a large expanding mass in the cubital fossa. Can you explain what has happened?

282 What is similar about the palmaris longus and peroneus tertius muscles?

D Forearm

Wrist joint

Information questions

283 What muscles attach at the common extensor origin?

284 Describe the wrist joint and its movements.

285 Abduction and adduction of the wrist are achieved by which muscles?

286 Why is it difficult to make a strong grip with a flexed wrist?

287 Describe the 'anatomical snuffbox'.

Applied questions

288 What surface anatomical clues aid clinical diagnosis of a fractured distal radius?

289 Why may diagnosing a median nerve injury not be as easy as it seems?

290 How can your sharp knowledge of upper limb vasculature variation save the hand of a child who has to undergo intravenous chemotherapy?

291 Which area of skin would be affected by (1) a C6 herpes zoster infection and (2) an axillary nerve injury (C5)?

292 What is the functional difference between injury to the ulnar nerve at the wrist and at the elbow?

293 An attempted suicide cuts his wrists fairly deeply. After control of the arterial bleeding, which other important structures would you wish to test and how would you do this?

294 A young woman falls on her outstretched hand and 6 weeks later complains of pain deep in the 'anatomical snuffbox' on palpation. Why should you X-ray her wrist?

295 You notice in 'cops and robbers' films that, to make a person drop a knife, the defender often hits the attacker's hand into acute flexion. Why is this a very sensible move?

E Hand

Information questions

296 Describe the arrangement of muscles and tendons attached to the middle finger.

297 Describe the flexor synovial sheaths.

298 What are the 'intrinsic' muscles of the hand?

299 Where are the dermatome and the myotome of T1?

300 Describe the nerve supply to the thenar and hypothenar muscles.

301 Describe the function of the interossei and lumbricals of the hand. Why do you need them to write properly?

302 How do the ulnar and the radial artery contribute to the blood supply of the hand? Give the surface markings of the palmar arches.

303 Describe which lymph node groups might become involved in an infection of the skin of the hand.

304 Describe the movements of the thumb. What is their relationship to those of the fingers?

Applied questions

305 What is 'cricket' (or 'mallet') finger?

306 An old woman falls downstairs and sustains a mid-shaft fracture of the humerus, damaging the radial nerve in the spiral groove. What are the effects, both motor and sensory, on the hand and wrist?

307 In the hand, which nerve lesion would be most serious, and why?

308 Why do the ring and little fingers have fixed flexion contractures in Dupuytren's disease?

309 In compression of the structures below the flexor retinaculum at the wrist (carpal tunnel syndrome), why is there weakness of the thenar muscles but normal sensation over the skin of the same region?

310 How might a swelling proximal to the wrist joint be connected with infection in the tip of the thumb?

311 A patient with leprosy has a wasted hand with 'guttering', a positive Froment's sign, and a claw of his ring and little fingers. What lesion is responsible?

312 Where would you perform an operation for Raynaud's disease of the hands? What might be its effects?

THORAX

A Thoracic wall

　Intercostal spaces

　Ribs

　Sternum

　Joints of the thorax

Information questions

313　Describe the commoner variations in sternal anatomy.

314　How does the manubriosternal joint vary with age?

315　Describe the actions and anatomy of the intercostal muscles.

316　What are the branches of a typical intercostal nerve?

317　Which are the atypical intercostal nerves?

318　Describe the blood supply to the intercostal spaces.

319　What are the surface markings of the sternal and costal lines of pleural reflection?

320　What is meant by the terms 'pump handle' and 'bucket handle' respiration?

321　Why is the vertebral level of T4 an important anatomical landmark?

322　What is meant by 'false' and 'floating' ribs?

323　Describe the articulations of the right sixth rib with the vertebral column.

Applied questions

324　Where may the urologist or general surgeon cause a pneumothorax?

325　Where do ribs fracture? What may be the consequences?

326 At which vertebral levels would one normally find the jugular notch, manubriosternal joint and inferior angle of the scapula?

327 A young Australian surfer presents with a small malignant melanoma on his back just medial to the inferior tip of the scapula. Which lymph nodes would you wish to examine or biopsy to detect any spread of the disease?

328 A paraumbilical tattoo, as practised by South American Indians, becomes infected. Which lymph nodes do you think will feel tender?

329 An anaesthetist performs a successful T5 nerve block. Why would there be no sensory loss on the anterior chest wall?

330 Why might a patient with tuberculosis of the T9 vertebral body present with abdominal pain?

331 A young man presents with shingles in a line just at the level of his umbilicus. Which branch of which spinal nerve is affected?

332 In coarctation of the aorta, the arterial blood finds a collateral circulation. Which vessels do you think are involved, and, after some time, what X-ray abnormalities might you see?

333 Following interchondral subluxation of the ninth costal cartilage, which muscles might go into spontaneous contraction?

334 Why does the first rib, on a chest X-ray, often look as if it ends in 'mid air'?

335 Why might a cervical rib cause pain in the region of the medial epicondyle of the humerus?

336 You are shown an X-ray of the sternum in which at least three separate sternebrae are visible. How old is the patient, approximately?

B Pleural cavity

Lungs

Information questions

337 What are the basic differences between the right and left lungs?

338 Describe the surface marking of the dome of the pleura.

339 Compare and contrast the nerve supply of the parietal and visceral layers of the pleura.

340 What is the pulmonary ligament?

341 What is meant by the term 'pleural cavity'?

342 Describe the lymphatic drainage of the lungs.

343 What are the costodiaphragmatic and costomediastinal recesses?

344 What is a bronchopulmonary segment?

345 Describe the position of the middle lobe of the right lung. To what does this lobe correspond on the left?

346 Describe the arterial supply and venous drainage of the main bronchi.

347 What is the lobule of the azygos vein?

Applied questions

348 Why does pleurisy often mimic other causes of upper abdominal pain?

349 If the pleura is ruptured, what are the immediate consequences?

350 You are informed that a patient has a pleural effusion. Where is the fluid situated?

351 How would you drain the pleural cavity of air, and through which structures would your needle penetrate?

352 Why might it be useful to know the surface marking of the dome of the pleura prior to neck vein catheterisation?

353 Why might an unlucky central venous catheterisation cause a chylothorax?

354 Where on a bronchogram might you see the apicoposterior segmental bronchus?

355 What altered anatomy gives diagnostic clues to the bronchoscopist?

356 You are told your patient has an apical basal pneumonia. Where on the chest would you expect to hear abnormal sounds with your stethoscope?

357 Your patient has unwanted bronchial secretions in her posterior basal segments. Into what position would she be best placed to aid postural drainage?

358 Why might a spreading apical carcinoma of the lung cause pain in the little finger and a partial drooping upper eyelid of the same side?

C Heart

Pericardium

Information questions

359 What is the 'base' of the heart?

360 Are the terms 'apex' and 'apex beat' synonymous?

361 Describe the surfaces of the heart.

362 What are the borders of the heart?

363 What is the infundibulum of the right ventricle?

364 What do the trabeculae carneae, moderator band and the supraventricular crest have in common?

365 Why is the left atrium poorly named?

366 Describe the embryological basis for the names of the aortic and pulmonary valve cusps.

367 Where are the aortic sinuses?

368 Describe the interventricular septum. Why are ventricular septal defects the commonest congenital heart defects?

369 Describe the features of the aortic valve.

370 What is the form and function of the 'cardiac skeleton'?

371 Describe the course and branches of the right coronary artery.

372 Discuss the variations in the anatomy of the coronary arteries.

373 Describe the conduction mechanism of the heart.

374 Describe the venous drainage of the heart.

375 Why are the coronary arteries so called?

376 Describe the course and distribution of the left coronary artery.

377 What is the function of the papillary muscles and chordae tendineae?

378 What vessels drain into the right atrium?

379 How many layers does the pericardium possess?

380 Describe the nerve supply to the parietal pericardium.

381 Where would you find the transverse and oblique sinuses of the pericardium?

382 What important structures are embedded in the fibrous pericardium?

Applied questions

383 Where do coronary arteries commonly occlude?

384 Why is the surgeon cautious during repair of an atrial septal defect?

385 What approaches exist for the emergency drainage of a cardiac tamponade?

386 What is the 'area of cardiac dullness'?

387 What is the surgical significance of the transverse pericardial sinus?

388 What are the surface markings of the heart valves, and why does one not listen at these sites to hear them best?

389 On a plain posteroanterior chest X-ray, what structures form the borders of the heart shadow?

390 Why is the pain of myocardial infarction not felt over the apex beat?

391 Why may one find an anatomically patent foramen ovale in a normal subject with no functional deficit?

392 During a knife fight your patient was stabbed medial to the apex beat in the fifth left intercostal space near to the sternum. Which chamber would you expect to have been lacerated?

393 Your patient has a heart block. What X-ray might reveal an abnormality related to the normal functioning of the atrioventricular node?

394 What use would a lateral view barium swallow X-ray be to a physician investigating a patient with mitral valve disease?

395 Why has the right atrium both smooth and rough areas?

D Anterior mediastinum

Posterior mediastinum

Superior mediastinum

Information questions

396 Describe the limits and contents of the anterior mediastinum.

397 Describe the origin, course and termination of the thoracic duct.

398 What is the origin of the splanchnic nerves, and what fibres do they contain?

399 What is the relationship between structures running vertically and those running horizontally in the posterior mediastinum?

400 What is the blood supply of the oesophagus?

401 What are the grey and white rami communicantes?

402 What is the stellate ganglion?

403 What are the contents of the superior mediastinum?

404 What is the ligamentum arteriosum?

405 Describe the origin and course of the left and right recurrent laryngeal nerves.

406 What drains into the left brachiocephalic vein? What are its relations?

407 Describe the common variations of the thoracic aorta.

408 Describe the course of the vagus nerve in the thorax.

409 What are the branches of the thoracic aorta?

410 Describe the azygos system of veins. What is their function?

Applied questions

411 What is the commonest oesophageal congenital defect?

412 During a barium swallow, what anatomical structures, in their normal or pathological states, may cause indentation of the oesophagus on an oblique X-ray?

413 What are the sentinel lymph nodes, and why are they more commonly enlarged on the left?

414 What is the danger of performing a tracheostomy in a child?

415 Which glandular organ may appear as a swelling in the anterior mediastinum, and what symptoms do you think it may cause?

416 Why do the trachea and the main bronchi not collapse during the negative pressure of inspiration?

417 Why might a patient with an apical carcinoma of the lung present with a hoarse voice?

418 Your patient has a carcinoma of the lower oesophagus. Where would you expect any metastases to spread?

419 What normal structures cause indentations of the anterior oesophageal wall as seen on a lateral barium swallow film?

420 Why should you be interested in the venous supply of the lower oesophagus when presented with an alcoholic Frenchman who has vomited 2 litres of blood?

421 Why should tracheo-oesophageal abnormalities be not infrequent in the newborn child?

422 At an exciting movie, a young boy throws a peanut into the air and, attempting to catch it in his mouth, inhales it. Into which bronchus is it likely to pass?

ABDOMEN

A Anterior abdominal wall

Inguinal canal

Testes and scrotum

Information questions

423 What do the linea alba, the linea semilunaris and the arcuate line of Douglas have in common?

424 What is the significance of the transpyloric plane?

425 Describe the form and function of the conjoint tendon.

426 What are the fascial layers of the anterior abdominal wall? How do they blend into the fasciae of the thigh and perineum?

427 Describe the contents of the inguinal canal and the coverings of the spermatic cord. What structure lies within the inguinal canal but outside the cord?

428 Describe the interplay between the gonads, the gubernaculum and the processus vaginalis in gonadal descent.

429 Where may you find the vestigial remnants of the embryonic genital ducts?

430 What are the horizontal lines on the rectus abdominis of a 'Chippendale' stage performer?

431 Describe the nine surface regions of the abdomen and how they are derived.

432 What is the tunica albuginea?

433 What single structure is easily palpable within the spermatic cord, and from where does it derive its blood supply?

434 Describe the course of a sperm from testes to urethra.

435 What is unusual about the superficial fascia of the scrotum?

Applied questions

436 What are the sites of abdominal wall herniae? Discuss the various types.

437 What is a gridiron incision, and what does it protect?

438 What are the common incisions of the anterior abdominal wall?

439 Where must you perform spinal anaesthesia for total scrotal anaesthesia?

440 Why should a person, after a paraumbilical tattoo, have swollen lymph nodes in both groins and both axillae?

441 Where is McBurney's point, and for which operation is it a landmark?

442 To drain off excess fluid within the peritoneal cavity (ascites), a trocar and cannula is used. Where would you insert it and through which structures would it pass?

443 Why should the rectus abdominis muscle be retracted laterally during a paramedian incision?

444 Describe the anatomical structures encountered during a Pfannenstiel incision.

445 A patient has a blocked inferior vena cava. How will blood return to the right atrium from the lower half of the body?

446 Why might your patient, following an inguinal hernia repair, complain of a pain in his scrotum?

447 Where does a kick in the testes hurt?

448 How and why does the lymphatic drainage of the scrotum and of the testes differ?

449 What are a hydrocele and a varicocele, and why is aspiration of a hydrocele often only of temporary relief?

450 To perform a vasectomy (deferential dochotomy), which layers of the spermatic cord must one incise?

451 What structures does the gubernaculum become in the adult female?

B Peritoneal cavity

Information questions

452 In which of the nine zones of the abdomen are the following to be found: stomach, spleen, liver and ileocaecal valve or orifice?

453 Which of the following structures are normally intraperitoneal: pancreas, spleen, kidney, liver, transverse colon and duodenum?

454 Describe the form and function of the greater omentum.

455 What is the difference between a fold, a ligament and a recess of the peritoneum?

456 What is the difference between *a* mesentery and *the* mesentery?

457 The omental foramen (Winslow) leads from the greater sac into what structure?

458 Describe the omental bursa and its recesses.

459 What is the falciform ligament and what structure lies within its free border?

460 Describe the position and form of the lesser omentum. What lies within its free border?

461 Where are the triangular and coronary ligaments?

462 What lies within the lienorenal ligament?

463 Who normally have openings into their peritoneal cavity?

464 What embryological structures do the median and medial umbilical ligaments represent?

Applied questions

465 What structures form within the ventral mesentery of the foregut?

466 What is ascites and, anatomically, where does it lie?

467 If your finger was in the lesser sac (omental bursa), what structures would be present posterior to it?

468 Why is the omental foramen (Winslow) very useful to the surgeon during a bloody cholecystectomy operation?

469 Why is a twisting of the bowel (volvulus) relatively common in the sigmoid colon but very rare in the descending colon?

470 Why is the pain of early appendicitis generalised and paraumbilical in site, whereas later on in the course of the disease it is localised to the right iliac fossa?

471 What are the subphrenic spaces, and why are they important?

C Gastrointestinal tract

Stomach

Small and large intestine

Information questions

472 Where may the vermiform appendix lie? How is it removed?

473 Discuss the location and surgical significance of an ileal diverticulum.

474 Where do the vagal trunks lie on entering the abdominal cavity?

475 Describe the innervation of the stomach.

476 Describe the structures lying in the 'stomach bed'.

477 Which major artery and its branches supply the foregut?

478 Describe the lymphatic drainage of the stomach.

479 How would you distinguish between jejunum and ileum?

480 How does the small intestine receive its autonomic nerve supply?

481 What are the distinguishing features of the large intestine compared to the small intestine?

482 Describe the arterial blood supply to the transverse colon.

483 Where is the origin of the sigmoid mesocolon?

Applied questions

484 What is congenital pyloric stenosis and when does it usually present?

485 A surgeon wishes to cut both vagus nerves (truncal vagotomy). Where do you think they are most easily found in the abdomen?

486 Why might a patient with a cancer of the stomach present with obstructed pancreatic drainage?

487 Under what circumstances might air appear in the peritoneal cavity, and how might this be seen radiologically?

488 What radiological features distinguish large from small bowel?

489 Why is the caecum sometimes found in the right hypochondrium?

490 Which region of the large bowel is most likely to suffer from ischaemia and why?

491 Diverticulosis of the colon is a common disease. Is there an anatomical explanation for the sites of development of this problem?

492 What principles underlie surgical colonic mobilisation?

493 Which parts of the colon are used for colostomies?

D Pancreas

Spleen

Duodenum

Information questions

494 Describe the embryology of the pancreas. Why is its duct system so variable?

495 What is the blood supply of the pancreas?

496 Why does the superior mesenteric artery appear to pierce the pancreas?

497 What is the relationship between the tail of the pancreas and the spleen?

498 Describe the four parts of the duodenum and their relation to the vertebral column and the head of the pancreas.

499 Describe the venous drainage of the spleen.

500 What do the splenic notches tell you about the spleen's development?

501 Describe the shape and surface markings of the spleen.

502 Describe the splenic artery.

Applied questions

503 How may gallstones cause pancreatitis?

504 How may pancreatic tumours cause ascites and jaundice? How may abnormal pancreatic anatomy cause vomiting?

505 What may be the consequence of a boy falling off a bike onto his left side? Why may the intra-abdominal injury sustained in this situation cause left shoulder tip pain?

506 Where may you find an accessory spleen? What is its importance?

507 How may either the biliary or the pancreatic ducts be viewed radiologically?

508 Why is the duodenum relatively difficult to mobilise during surgery?

E Liver

Biliary system

Portal venous system

Information questions

509 How may the right hepatic artery vary?

510 Name the lobes of the liver. Are they functionally distinct?

511 Describe the blood supply of the liver.

512 What is the 'bare area' of the liver? Describe its boundaries.

513 Can you expect to feel the liver in a normal person?

514 Describe, in the fetus, the path taken by blood from the umbilicus to the inferior vena cava, and how this changes after birth.

515 What structures lie in the porta hepatis?

516 Describe the extrahepatic biliary tree.

517 Describe the most common arrangement in the formation of the portal vein.

518 Where are the main sites of portosystemic venous anastomoses?

Applied questions

519 In performing a liver biopsy, why should pneumothorax be not an uncommon complication?

520 Why must the patient hold his breath during a liver biopsy?

521 What is Riedel's lobe and why is it worth knowing about?

522 By what technique could you view the venous drainage of the liver?

523 How can you explain obstruction of the bowel due to gallstones?

524 Inflammation of the gall bladder may present as pain felt in which region?

525 Why might a patient with a long history of gallstones suddenly become jaundiced?

526 Where is pain from biliary colic felt?

527 Discuss the common variations in the anatomy of the biliary system that account for most of the errors in gall bladder surgery.

F Kidney

Ureter

Suprarenal gland

Information questions

528 Describe the anterior and posterior relations of the kidneys.

529 Describe the 'capsules' of the kidney.

530 A pelvic kidney may derive its blood supply from which artery?

531 What major difference is there between the tributaries of the right and left renal veins?

532 Describe the normal flow of urine from a minor calyx to the bladder.

533 Describe the blood supply to the ureters.

534 Describe the relationship of structures at the hilum of the kidney.

535 Describe the relations of the suprarenal glands.

536 What is the connection between the suprarenal cortex and the sympathetic nervous system?

Applied questions

537 How could you explain the occurrence of a left-sided varicocele in a man with a palpable abdominal mass and blood in his urine (haematuria)?

538 Describe the structures encountered in performing a lower pole renal biopsy.

539 What are the surface markings of the kidneys, and what explains their difference in position?

540 Why might a patient have a pneumothorax after a renal biopsy?

541 Why should a horseshoe kidney be situated at a lower level than the normal organ?

542 How might the upper urinary tract be viewed radiologically?

543 What are the classic markings of the right ureter on an abdominal X-ray?

544 What structures may be damaged in the common surgical approach to the kidney?

545 Where is ureteric colic felt?

546 Explain the location of ectopic kidneys. How may they prove dangerous?

547 Which relatively narrow sites of the ureter would you check for the impaction of stones?

G Posterior abdominal wall

Diaphragm

Information questions

548 Describe the relationship of the psoas major muscle and the branches of the lumbar plexus.

549 Name the paired arterial branches of the abdominal aorta.

550 What are the tributaries of the inferior vena cava, and what makes them asymmetrical?

551 Describe the attachments of the iliac and thoracolumbar fasciae.

552 Describe the relations of the inferior vena cava.

553 What are the tributaries of the inferior vena cava?

554 How may blood return to the heart if the inferior vena cava is obstructed?

555 What is a persisting left inferior vena cava?

556 What is the cisterna chyli and how may it be visualised?

557 What is the hypogastric plexus?

558 Describe the lymph nodes of the posterior abdominal wall.

559 What is the relationship of the common iliac artery, ureter and testicular vessels at the brim of the pelvis?

560 Explain the relationships to each other of the following structures: ureter, testicular vessels, psoas muscle and right colic artery.

561 What are the origins of the diaphragm?

562 What are the median, medial and lateral arcuate ligaments?

563 From which main structures is the diaphragm formed and how does this explain its motor and sensory nerve supply?

564 Which structures pass through the diaphragm and at what approximate vertebral levels?

Applied questions

565 Discuss the clinical significance of the relations of the iliopsoas muscle.

566 What is the solar plexus, and if you receive a blow to this region what happens?

567 A young Indian boy complains of a swelling in his lumbar spine and, more recently, a swelling in the groin. Anatomically, how can you connect these two symptoms?

568 Describe the surgical approach to a lumbar sympathectomy. How may the unfortunate surgeon amuse the pathologist?

569 On which X-ray examination would you hope to find the cisterna chyli, and what is its function?

570 Describe the surface markings of the abdominal aorta.

571 In a patient with a testicular tumour, where would you wish to examine lymph nodes for suspected metastases?

572 Discuss the likely sites of both congenital and acquired herniae of the diaphragm.

573 Where do you think pain is felt after an injury to the central portion of the diaphragm?

574 Through which part of the diaphragm does the inferior vena cava pass, and why must this be so?

575 What neurosurgical procedure can damage the inferior vena cava?

PELVIS AND PERINEUM

A Bones and ligaments

Sexual differences

Information questions

576 Why is the sacroiliac joint classified as an atypical synovial joint?

577 What factors stabilise the sacroiliac joint?

578 What structures cross the alar of the sacrum?

579 Define the true and the false pelvis.

580 Given a bony pelvis, how would you position it correctly?

581 What are the pelvic inlet (brim) and the pelvic outlet?

582 What type of joints are the pubic symphysis and sacroiliac joints?

583 What is the relationship between the lacunar and inguinal ligaments?

584 How might the greater sciatic notch help you distinguish a male from a female pelvis?

585 What creates the difference between the male and the female bony pelvis?

Applied questions

586 How may the pelvis be fractured?

587 What bony landmarks do you think are used when performing a pudendal nerve block during labour?

588 What are the true and diagonal conjugates?

589 Why might the sacral hiatus prove a very useful landmark for the anaesthetist?

590 The 'dimples of Venus' are easily seen when watching young women in bikinis! What are they, and for what anatomical landmarks are they a useful guide?

591 On vaginal examination, you find a woman to have a subpubic angle of about 60°. What does this tell you about her pelvis?

592 On vaginal examination in early pregnancy, you estimate your patient's ischial spines to be 8 cm apart. As her obstetrician, about what might you worry?

593 What are the normal measurements of the pelvic brim in a woman?

B Perineum

Male and female urogenital regions

Information questions

594 With what is Colles' fascia of the perineum continuous over the anterior abdominal wall?

595 Describe the main branches of the pudendal nerve.

596 Describe the three divisions of the male urethra and the two which are within the perineum.

597 What structures lie within the deep perineal pouch or space?

598 What muscles lie within the superficial perineal space or pouch?

599 Though artificial, what are the boundaries of the urogenital triangle?

600 Why is the dorsum of the penis so described?

601 Discuss the anatomy of penile erection.

602 What is the perineal body and to which structures is it attached?

603 Where would you find the bulbourethral glands (Cowper) and what is their female homologue?

604 Describe the areas of drainage to the superficial inguinal lymph nodes.

Applied questions

605 How would you identify the site of the perineal membrane on a male urethrogram?

606 On a pelvic examination, where would you find an inflamed greater vestibular gland (Bartholin)?

607 Which are the widest and narrowest parts of the male urethra?

608 After rectal surgery, why are some men unable to ejaculate?

609 What do you think might be the consequences of a ruptured perineal body?

610 What structures are cut in a mediolateral episiotomy?

611 What nerves must be blocked in perineal anaesthesia during childbirth?

612 What is hypospadias, and how does its occurrence help explain the development of the penile urethra in the male?

613 What anatomical knowledge is essential before urethral instrumentation is performed?

614 What factors influence the course of extravasated urine after a boy ruptures his urethra falling astride the crossbar of his bicycle?

615 What is circumcision, and why should special care be taken when performing surgery around the frenulum?

C Pelvic floor

Rectum

Anal triangle

Information questions

616 Discuss the form and nerve supply of the pelvic diaphragm in both sexes.

617 Discuss the functions of the pelvic diaphragm.

618 Describe the boundaries and contents of the ischioanal fossa.

619 What are the internal and external anal sphincters and how are they innervated?

620 Describe the blood supply of the rectum and anus.

621 What are the rectal valves or transverse folds?

Applied questions

622 What part of the pelvic diaphragm is most likely to be damaged in childbirth?

623 What have the pectinate line and hymen in common?

624 The pectinate line demarcates the proctodeum and cloaca of the embryo. How is this relevant to understanding metastases from an anal carcinoma?

625 What is the embryological explanation of an imperforate anus?

626 What may be felt during a digital rectal examination in both sexes?

627 Why may haemorrhoids be associated with cirrhosis of the liver, and, if injected in the correct region, cause no pain?

628 What is the lithotomy position? Describe the boundaries of the anal triangle in this position.

629 What is the major content of the ischioanal fossa, and why is it often a site of abscess formation?

630 How may ovarian cancer and an obturator hernia commonly present?

D Male pelvic organs

Information questions

631 Into where do the common ejaculatory ducts open?

632 Describe the course and blood supply of the ductus deferens.

633 What are the seminal vesicles?

634 Discuss the innervation of the bladder and associated urogenital structures.

635 What are the branches of the internal iliac artery?

636 Describe the capsules of the prostate gland.

637 Describe the peritoneal coverings of the rectum.

638 Describe the features of the prostatic urethra. What is the 'male vagina'?

639 Describe the boundaries of the bladder trigone. What type of receptors are particularly numerous in this region?

640 What is the verumontanum?

Applied questions

641 What remnants of the paramesonephric ducts are there in the adult male?

642 What is the clinical significance of drainage from the prostatic venous plexus?

643 Why might an enlarged prostate gland cause retention of urine?

644 What structures are normally visible on a vasogram X-ray?

645 A man complains of passing blood in his urine (haematuria) and, at the end of micturition, a very severe pain at the tip of the penis. With regard to the anatomy of the bladder, what comments can you make?

646 Why does the suprapubic catheterisation of a full bladder not result in peritonitis?

647 In which normal people might one expect to percuss or palpate a bladder?

648 What is a deferential dochotomy?

649 What factors alter the normal palpable anatomy of the prostate gland?

650 How does the anatomy of the pelvis aid the colorectal surgeon?

651 Discuss the anatomy of ejaculation. What is the anatomical basis for impotence after colorectal surgery in the male?

E Female pelvic organs

Information questions

652 From where does the ovary receive its blood and nerve supply?

653 What is the homologue of the round ligament of the uterus in the male embryo?

654 What is the female equivalent of the prostate gland?

655 What is a retroflexed, retroverted uterus?

656 Describe the uterine tubes and their relationship to the broad ligament.

657 Describe the lymphatic drainage from the uterus.

658 What are the embryological urachus and the obliterated umbilical artery called in the adult?

659 What are the features and contents of the broad ligament of the uterus?

660 What are the fornices of the vagina?

661 Describe the peritoneum of the female pelvis in contrast to that of the male.

662 What supports the uterus?

Applied questions

663 Why do gynaecologists have nightmares about ureters?

664 Where may calculi obstruct the ureters?

665 During what examination might one, by chance, feel pathology of the ureters?

666 What structures can one normally palpate during a bimanual pelvic examination?

667 During a hysterosalpingogram, contrast medium is seen to flow into the rectouterine pouch (Douglas). What does this tell you about the woman's anatomy?

668 Name three structures – all remnants of the mesonephric ducts – which may, in the adult female, give rise to cystic swellings.

669 Why may an amateur abortionist accidentally kill his or her patients, due to peritonitis?

670 Why do some women experience paraumbilical pain about 14 days prior to menstruation?

671 Can you explain an abdominal pregnancy?

672 What consequences might follow stretching of the cardinal and uterosacral ligaments?

673 To which lymph nodes would an ovarian tumour most likely metastasise?

LOWER LIMB

A Gluteal region

Hip joint

Information questions

674 What bursae are associated with the gluteus maximus muscle?

675 Describe how the capsule of the hip joint is adapted for its various functions.

676 Describe the ligaments of the hip joint and their functions.

677 What are the 'guy ropes' of the hip bone?

678 What converts the greater and lesser sciatic notches into foramina?

679 Describe the structures which exit from the greater sciatic foramen below the piriformis muscle.

680 What structures enter the lesser sciatic foramen from the greater sciatic foramen?

681 Describe the course of the sciatic nerve as far as its bifurcation into common peroneal and tibial branches.

682 Describe the ligaments of the hip joint. Why is hip extension limited to only about 15°?

683 Describe the muscles involved in abduction of the hip. What is their most important function in everyday life?

684 Which are the six small lateral rotator muscles of the hip?

Applied questions

685 Discuss the anatomical basis for the 'gluteal gait'.

686 Discuss the anatomical consequences of hip dislocation.

687 When performing intramuscular injections into the gluteal region, what anatomical landmarks would you use to make sure you did not damage the sciatic nerve?

688 A patient with prolapsed intervertebral discs at L4/5 and L5/S1 complains of sciatica. Where will the pain be worst, and which dermatomes will you examine to test for loss of sensation?

689 Why should a subcapital fracture in the elderly often lead to avascular necrosis of the femoral head, whilst a pertrochanteric fracture usually heals well with a pin and plate procedure?

690 Why, on examination of a patient with a fractured neck of femur, would you often find shortening of the limb and lateral rotation of the foot?

691 Why does it seem that a ballet dancer can extend the hip to 90°? What, in fact, is she doing?

B Femoral triangle

Adductor compartment of thigh

Information questions

692 Discuss the cutaneous innervation of the thigh.

693 Describe the form and function of the fascia lata.

694 What is the saphenous opening?

695 Describe the course and surface marking of the femoral artery.

696 What is the importance of the cruciate anastomosis?

697 What is the anatomical significance of enlarged superficial inguinal lymph nodes?

698 Describe the quadriceps femoris muscle.

699 Describe the boundaries, roof, floor and contents of the femoral triangle.

700 Describe the adductor muscles and their innervation.

701 Describe the adductor canal (Hunter's, or the subsartorial canal) and its contents.

Applied questions

702 On examination, you find your patient has a small femoral hernia which is reducible. After reduction, which is often not possible in this type of hernia, your finger lies in the femoral canal. What structures lie deep, superficial, medial and lateral to your examining finger?

703 Where and how would you find the femoral vein to obtain venous blood in a collapsed patient?

704 By observation alone, how could you tell the difference between a femoral and an indirect inguinal hernia?

705 The great saphenous vein is homologous to which vein of the arm?

706 A patient with an ulcer on the glans penis should have which lymph node group examined?

707 Why should a patient with a tumour within the psoas muscle complain of tingling in his big toe and numbness of the side of his buttock and thigh?

50

708 Why might a patient complain of pain and paraesthesia in the medial side of the lower leg following stripping of the great saphenous vein for varicosities?

709 A butcher's knife slips and plunges into his thigh at the apex of the femoral triangle. Why is this likely to be particularly bloody?

C Knee joint

Popliteal fossa

Information questions

710 Describe the menisci of the knee joint. What is their function?

711 Why does the patella tend to dislocate laterally? How is this prevented?

712 Describe the bursae around the knee joint. Which are in direct communication with the joint?

713 What is the function of the cruciate ligaments, and what happens to them in flexion and hyperextension?

714 Which hamstrings have an action on the knee joint?

715 Describe the muscles and nerves involved in the knee jerk.

716 Describe the boundaries, floor, roof and contents of the popliteal fossa.

717 Describe the functional importance of the popliteus muscle.

Applied questions

718 Why does the knee feel unsteady after prolonged bed rest or after a resolved effusion?

719 What anatomical factors contribute to fracture of the patella?

720 Why is a knowledge of the suprapatellar bursa clinically important?

721 What factor is most important to the stability of the knee joint?

722 An 8-year-old boy comes into your surgery with a pain in the knee. Why would you examine thoroughly both his knee and his hip?

723 Footballers often have to undergo operations for removal of one or both menisci. Which is more commonly torn and why?

724 Why can a physician obtain a 'patella tap' on a knee joint containing an effusion?

725 What are 'housemaid's knee' and 'clergyman's knee'?

726 Why should women suffer from dislocation of the patella more commonly than men? Name three factors which normally prevent lateral dislocation.

727 Describe the development of the patella. How might this explain the appearance of what looks like a fracture in a normal person?

728 A partly decomposed skeleton of a baby is found. Why might an X-ray of the knee be an important piece of evidence in distinguishing whether the child had been stillborn or had died after a short extrauterine life?

729 Why might a 'locked' knee be the presenting sign of a torn meniscus?

730 Why is a popliteal pulse difficult to feel even in the normal individual?

731 What structures do you think may be the cause of a swelling in the region of the popliteal fossa?

732 Which tendons are easily palpable at the back of the knee?

733 A young man is crossing the road and is knocked down by a bus. On arriving at the hospital it is noticed that he has a very swollen lateral side of the knee and lower leg and that he cannot dorsiflex his foot. Can you explain his 'foot-drop'?

D Compartments of the leg

Ankle joint

Information questions

734 Which muscles are found in the flexor and the extensor compartment of the leg? Which muscles constitute a separate, third, compartment?

735 Describe the muscles involved in plantar flexion and in dorsiflexion.

736 Describe the superficial venous drainage of the lower limb.

737 Describe the tendons in the extensor compartment at the ankle joint.

738 What are the nerve supply and particular action of the peronei muscles?

739 Describe the retinacula. What is their function?

740 Describe the major ligaments of the ankle joint.

741 Describe the muscles and nerves involved in the ankle jerk.

Applied questions

742 What structures may cause acute calf pain when damaged?

743 What are 'shin splints'?

744 How would you test separately the functions of soleus from gastrocnemius?

745 Why might stripping the short saphenous vein for varicose veins leave the patient with tingling in his little toe?

746 How would you display radiologically the veins of the leg? Why might this be useful clinically?

747 You are called to an emergency where a patient requires an instant intravenous line. All his veins are collapsed so you take a scalpel and 'cut-down'

on the great saphenous vein. Where would you hope to find it?

748 Who do you think sustains a rupture of the tendo calcaneus most often? What do you think are the consequences?

749 Does everyone have both the dorsalis pedis and posterior tibial pulses present?

750 Why are ankles sprained more commonly whilst walking downstairs than when going upstairs?

751 In a sprained ankle that has a forced eversion injury, the deltoid ligament may tear. Why is this quite a serious situation?

752 What is a Pott's fracture of the ankle, and why do you think a third degree type may often require internal fixation?

E Foot

Posture and walking

Information questions

753 What are the 'extra' foot bones?

754 Describe the four layers in the sole of the foot.

755 Discuss the joints, ligaments, muscles and movements involved in inversion and eversion of the foot.

756 What anatomical structures are responsible for creating the imprint of a wet foot on the bathroom floor?

757 What factors maintain the integrity of the arches of the foot?

758 Describe the four named peripheral nerves which supply sensation to the dorsum of the foot.

759 Name the factors which maintain the arches of the foot.

760 Where is the line of gravity?

761 Describe the major muscle groups involved in normal walking, and their sequence of contractions.

762 An infected cut on the heel will cause tenderness in which lymph nodes?

Applied questions

763 Describe the common fractures of the foot.

764 Which arteries may be severed by standing on a broken glass bottle?

765 Discuss the anatomical basis for 'flat feet'.

766 In a severe sprain of the ankle, why is it not uncommon to find that the patient has an avulsion of the fifth metatarsal tuberosity?

767 What is talipes equinovarus, and in whom may it be commonly seen?

768 In flat foot (pes planus), which arch is particularly depressed?

769 Why should a patient, following poliomyelitis of the lower leg, present with claw-foot (pes cavus)?

770 Why might a lumbar sympathectomy help a patient with occlusive arterial disease of the lower limb?

BACK

A Vertebral column

Joints and ligaments

Information questions

771 Discuss the characteristic features of typical cervical, thoracic and lumbar vertebrae.

772 Which cervical spinous process is the most easily palpated?

773 Where is the body of the atlas?

774 At what vertebral level do the iliac crests usually lie?

775 Why is movement in the lumbar region mainly flexion and extension?

776 What type of joint lies between vertebral bodies?

777 Describe the specialised ligaments of the atlanto-axial joint.

778 Which animals have a very strong ligamentum nuchae, and from what is it formed?

779 Describe the main ligaments attached to the body and spine of a thoracic vertebra.

Applied questions

780 What are kyphosis, scoliosis and lordosis?

781 What is spina bifida?

782 Why do you think the posterior superior iliac spines are a useful landmark for a doctor wishing to examine the bone marrow?

783 On an oblique X-ray of the lumbar vertebrae, what is the 'Scottie dog'?

784 Why should the cervical vertebrae be prone to dislocation in whiplash injuries?

785 What might be an immediate consequence of a fracture dislocation of the dens?

786 Why would direct injection of contrast medium into the vertebral artery above C6 prove extremely difficult?

787 Why might a sudden posterior prolapse of the nucleus pulposus between C2 and C3 prove fatal?

788 What structures are pierced on performing a cisternal (cerebellomedullary cistern) puncture?

789 Which is the most frequent site of herniation of the nucleus pulposus, and what may be the consequences?

B Soft tissues of the back

Suboccipital region

Information questions

790 What are the main muscle groups forming the erector spinae?

791 Of which muscles is the transversospinalis group composed?

792 What is the nerve supply to all the deep back muscles?

793 What is unusual about the dorsal ramus of C1 (suboccipital nerve)?

794 Describe the contribution made by the ventral rami to the sensory supply at the back of the neck.

795 Describe the small muscles involved in atlanto-occipital and atlantoaxial movements.

796 What is the most vital structure in the suboccipital triangle?

797 What forms the roof and floor of the suboccipital triangle?

Applied questions

798 Where are the triangle of auscultation and the lumbar triangle?

799 Why is the vertebral venous plexus (Batson) of importance in understanding the spread of prostatic malignancy?

800 Why might a left-sided leg pain (sciatica) and a low backache be associated with spasm of the left lower erector spinae?

801 If a patient complained of pain on the outer side of the upper arm, which increased in intensity during neck movement, which cervical vertebrae would you wish to X-ray?

802 On a cervical spine X-ray, what might you look for in the soft tissue region of a patient with tingling in both forearms whilst carrying suitcases?

C Spinal meninges

Epidural space

Information questions

803 Describe the denticulate ligaments.

804 What is the filum terminale and where does it terminate?

805 Where do the spinal cord and spinal theca normally terminate in an adult?

806 How many pairs of nerves are derived from the spinal cord?

807 What lies within the subarachnoid space?

808 What are the contents of the epidural (extradural) space in the normal adult?

809 How far laterally does the dural sheath extend?

Applied questions

810 On a myelogram, why does one often see contrast medium around the nerve roots?

811 Why is a lumbar puncture at the level of L2 potentially hazardous in an infant?

812 Why does the normal lateral erect myelogram have indentations on its anterior thecal margin, and for what reason are these checked?

813 The intervertebral foramen has bony and cartilaginous walls. Why is this of significance?

814 What structures would be encountered in performing a lumbar puncture just below the spine of L4?

Answers

HEAD AND NECK

A Information answers

1   The bones of the vault of the skull, being flat, are ossified in membrane. Those of the base of the skull are, in general, ossified in cartilage. Thus the frontal bone, parietal bone, the squamous parts of the occipital and temporal bones and the greater wings of the sphenoid are ossified in membrane, as are the facial bones, viz. maxilla, zygomatic, palatine, lacrimal bones and the vomer.

2   The temporal bone is in four parts at birth: the squamous, styloid process, petrous and tympanic parts. The petrous part and styloid process are developed in cartilage, the other two in membrane.

3   The frontal bone ossifies in membrane from two centres, one on each side. These two are separated at birth by a suture – the metopic suture – situated in the median sagittal plane.

4   The vertex is the highest point of the skull in the anatomical position. It is situated in the median sagittal plane between the two parietal bones. The inion is synonymous with the external occipital protuberance. This is a bony protuberance of the occipital bone in the midline, for the attachment of the ligamentum nuchae. Above each orbit is the supraciliary ridge which is more prominent in males. The two supraciliary ridges are joined across the midline by an elevation – the glabella. The crista galli is situated on the interior of the skull. It is a sharp upward projection of the ethmoid bone in the midline of the anterior cranial fossa. It separates the cribriform plates and gives attachment to the falx cerebri.

5   Immediately medial to the spine of the sphenoid is situated the junction between the bony and

cartilaginous portions of the auditory tube (Eustachian, pharyngotympanic), which the spine therefore protects. Anteromedially is situated the foramen spinosum, for the middle meningeal artery, which is therefore also protected to a degree. A further important relationship is the opening of the carotid canal posteromedially.

6   The sphenoid usually has six apertures through it. The optic canal lies between two roots of bone which join the body to the lesser wing. It connects the cranial cavity with the orbit and transmits the optic nerve and ophthalmic artery. Four further foramina lie in a semicircle on the medial aspect of the middle cranial fossa. Most anteriorly is the superior orbital fissure, connecting the fossa to the orbit and conducting the branches of the ophthalmic division of the trigeminal nerve (V), the oculomotor (III), trochlea (IV) and abducent (VI) nerves, and superior ophthalmic vein. Behind the superior orbital fissure is the foramen rotundum. It conducts the maxillary division of the trigeminal nerve (V) on its way to the pterygopalatine fossa. Next is the foramen ovale, which transmits the mandibular division of the trigeminal nerve (V) (both sensory and motor parts) and the lesser petrosal nerve. Most posterior is the foramen spinosum through which the middle meningeal vessels enter the skull. The pterygoid canal links the pterygopalatine fossa with the foramen lacerum. It is situated at the base of the pterygoid process. It conducts the nerve of the pterygoid canal containing both sympathetic and parasympathetic elements on their way to the pterygopalatine ganglion.

7   The diploic veins drain the diploic space. This is the marrow-containing portion of cancellous bone situated between the inner and outer layers of compact bone, in the bones of the vault of the skull. The diploic veins emerge to the exterior of the skull. There are usually four on each side: one frontal, two parietal and one occipital.

8   The emissary veins connect the intracranial venous sinuses with veins on the outside of the skull. They are valveless and therefore blood may flow in either direction in them. They are clinically important because they may transmit extracranial infection to the inside of the skull.

9 The conchae are projections of the lateral wall of the nose into the nasal cavity. They consist of a bony skeleton covered in mucous membrane. There are three conchae on either side, named superior, middle and inferior. The superior and middle conchae are derived from the ethmoid bone; the inferior is a separate bone in its own right.

10 The foramen lacerum is situated in the middle cranial fossa between the irregular apex of the petrous temporal bone and the sphenoid. As its name suggests, it is irregular in shape. In life it is filled with cartilage and fibrous tissue. Only a few small veins actually pass through it but the internal carotid artery passes directly over it during its tortuous passage into the skull.

11 The word 'antrum' is frequently used to describe the maxillary air sinus. This is a pyramidal air-filled space located in the body of the maxilla and lined with respiratory epithelium. It communicates with the middle meatus of the nasal cavity via the hiatus semilunaris.

12 The joint between the basi-occiput and the basisphenoid is a cartilaginous joint of the synchondrosis variety. The two bones' growing edges are separated by a block of cartilage into which they both grow and eventually fuse. This joint resembles, therefore, the joints between long bones and their epiphyses. The cartilage of the basi-occiput/basisphenoid joint is obliterated and bony union occurs around the twenty-fifth year.

A Applied answers

13 The most obvious differences are the open sutures and fontanelles which allow moulding of the skull during delivery. The anterior fontanelle (bregma) is diamond-shaped and bounded by the developing frontal and parietal bones. The frontal bones normally fuse, but occasionally persist in the adult as the metopic suture. The posterior fontanelle (lambda) is triangular-shaped and lies between the occipital and parietal bones. These two fontanelles close by 18 months and 6 months, respectively.

The general bony proportions are different, the facial skeleton being relatively tiny at birth and only enlarging during the first few years. The alveolar processes and paranasal air sinuses are only rudimentary at birth and expand during growth, as do the mastoid air cells. These differential growth rates contribute to internal changes such as the shape and position of the auditory tubes as well as the more obvious changes such as the rounded baby face becoming an elongated adult face. Less obvious differences include the structure of the external auditory meatus. In the newborn its walls are cartilaginous as far as the tympanic membrane, whereas in the adult the medial part of the meatus is part of the temporal bone and is therefore bony.

14 The mastoid air cells do not develop until the second or third year, and consequently the relatively small size of the newborn mastoid leaves the facial nerve exposed as it exits the skull through the stylomastoid foramen. A superficially placed nerve is therefore easily damaged, especially during forceps delivery.

15 The normal beard overlies the zygomatic portion of the temporal bone, zygomatic bone, the ramus, angle and body of the mandible and the maxilla. An extremely long beard may of course overlie the manubrium and body of the sternum!

16 The pterion is the meeting point of temporal, parietal and frontal bones with the greater wing of the sphenoid bone. In the fetal skull it may be seen as a fontanelle, whereas in the adult it lies some 4 cm above the midpoint of the zygomatic arch. Immediately deep to this point lies the middle meningeal artery and, more specifically, its anterior division, which may be lacerated in a direct blow causing a fracture injury to the side of the head. The bone is relatively thin in this area and an extradural haematoma may follow a direct hit with a missile such as a golf or cricket ball.

17 They are hollow bones containing air-filled spaces. The mastoid air cells lead into the middle ear whilst the ethmoid, sphenoid and maxillary paranasal air sinuses all open into the lateral wall of the nasal cavity. These air spaces are said to lighten the weight of the skull and resonate the voice.

18 The hyoid bone is a U-shaped structure with a body and four horns (cornua). It is the upper-most platform of the laryngeal cartilages and is derived from the second and third arch cartilages. It is suspended from the styloid process by stylohyoid ligaments and has attached to it some dozen or so small muscles. These connect the hyoid bone to the tongue, jaw, sternum, larynx and pharynx. Consequently, following hyoid trauma the fine balance of these muscles would be disturbed, causing problems in breathing, speaking and swallowing.

19 A fracture of the mandible may easily involve the inferior alveolar (dental) nerve which lies within the bone. Its terminal branches exit the bone through the mental foramen where they supply the mucous membrane of the lower lip and chin. (Remember what happens to your lip following an inferior alveolar nerve block whilst undergoing dental procedures!)

B Information answers

20 The anterior cranial fossa is situated anterior to the lesser wings of the sphenoid bone. The middle cranial fossa lies between the lesser wing of the sphenoid anteriorly and the two petrous temporal bones posteriorly. The frontal lobes of the brain occupy the anterior cranial fossa, the temporal lobes the middle cranial fossa.

21 The clinoid processes form the margins of the sella turcica. They provide attachments for the tentorium cerebelli – the attached border being attached to the posterior clinoid processes, the free border being attached to the anterior processes. Between the processes is stretched the diaphragma sellae, the dural roof of the sella turcica. Within the sella lies the pituitary gland.

22 The clivus is the inclined surface between the anterior part of the foramen magnum and the sella turcica. It is composed of the occipital bone posteroinferiorly and the sphenoid anterosuperiorly, with the synchondrosis between them. The medulla lies against the lower aspect of the clivus, the pons against the upper part.

23 The jugular foramen lies in the posterior cranial fossa between the lower border of the petrous temporal bone and the occipital bone. It transmits the sigmoid sinus, which becomes the internal jugular vein on passing through the foramen. The inferior petrosal sinus descends in a groove between the petrous temporal and occipital bones, to enter the foramen. It passes through the foramen anterior to the sigmoid sinus, to enter the internal jugular vein. The IXth, Xth and XIth cranial nerves also leave the skull through the jugular foramen. The IXth nerve grooves the inferior border of the petrous temporal bone as it does so. These three nerves exit anterior to the sigmoid sinus. From anterior to posterior, they are arranged IX, X, XI.

24 In the neck, the hypoglossal nerve (XII) descends in the carotid sheath anterior to the vagus. At the lower border of the posterior belly of digastric, it loops around the occipital artery and, turning anteriorly, crosses superficial to both carotid arteries. As it progresses, it crosses the loop of the lingual artery superficially, and comes to lie on the superficial aspect of the hyoglossus muscle. It is unique in having this superficial relationship to the carotid arteries.

25 In certain areas the arachnoid projects into the intracranial venous sinuses to form arachnoid granulations. These granulations serve to drain cerebrospinal fluid (CSF) into the venous blood. They are particularly numerous around the superior sagittal sinus where they groove the skull, forming pits in the internal aspect. The superior sagittal sinus has lateral extensions in various places. These are termed lacunae. They, too, contain arachnoid granulations.

26 The falx cerebri is a sickle-shaped double fold of the inner dura which projects into the longitudinal fissure in the midline between the cerebral hemispheres. Anteriorly, it is narrow and is attached to the crista galli. Posteriorly, it is broad and supports the tentorium cerebelli, as a ridge would a tent. In this posterior region, the straight sinus is found. The fixed margin of the falx cerebri, attached to the skull in the median plane, contains the superior sagittal sinus between its layers. Its free inferior margin contains the inferior sagittal sinus. The falx

cerebri protects the brain by preventing excessive side-to-side displacement of the cerebral hemispheres. Because it contains the sinuses, it has a role in the venous drainage of the brain.

The falx cerebelli is a small double fold of the inner dural layer. It lies in the mid-sagittal plane, being attached to the internal occipital crest. Superiorly it is attached to the inferior aspect of the tentorium cerebelli. Its anterior free margin fits into the posterior cerebellar notch. Its function is to help maintain the cerebellum in place.

27 The cavernous sinuses lie on either side of the body of the sphenoid bone. They extend from the superior orbital fissure, anteriorly, to the apex of the petrous temporal bone, posteriorly. Numerous fibrous strands, covered with endothelium, cross their lumina. The internal carotid artery passes through the sinus from posterior to anterior, hugging its medial wall. Its adventitia is separated from the venous blood of the sinus by a layer of endothelium. The artery is surrounded by a plexus of sympathetic nerves. The abducent nerve (VI) passes through the sinus inferior to the internal carotid artery. The lateral wall of the sinus has four nerves embedded in it. From superior to inferior these are the oculomotor (III), the trochlea (IV), the ophthalmic division of the trigeminal (V_1) and the maxillary division of the trigeminal (V_2). Posteriorly the trigeminal ganglion replaces these last two nerves. It lies in a dural cave (Meckel) on the lateral wall of the sinus.

28 The afferent veins of the cavernous sinus are three: first, the superior ophthalmic vein, via the superior orbital fissure, drains the orbit into the anterior part of the cavernous sinus; second, the middle cerebral vein pierces the roof of the sinus; third, the sphenoparietal sinus runs round the lesser wing of the sphenoid and pierces the roof of the sinus. Also draining into the sinus are the inferior ophthalmic vein and the central vein of the retina. The efferent veins of the cavernous sinus are the superior petrosal sinus and the inferior petrosal sinus. The sinus also communicates with the pterygoid plexus of veins via the foramen ovale. It should be noted that the cavernous sinuses communicate with each other across the midline via the intercavernous sinus.

29 In the child of 6 months the bones of the vault of the skull are not united by sutures. They are separated by fibrous membranes and, at the corners of the bones, these membranes form large areas called fontanelles. The anterior fontanelle is diamond-shaped and is situated between the two frontal and the two parietal bones. It disappears at about the age of 18 months. The posterior fontanelle is triangular. It lies between the occipital bone, posteriorly, and the two parietal bones, anterolaterally. It is fused by the end of the first year. The fontanelles may be palpated as soft depressions. Together with the interparietal suture, they resemble an arrow, pointing anteriorly. This has led to the adoption of the word sagittal (arrow-like) to describe this particular anatomical plane. Furthermore, the superior sagittal sinus lies immediately inferior to these fontanelles in the mid-sagittal plane. A needle may be passed through its fibrous roof into this sinus, therefore, for the withdrawal of venous blood from an infant.

30 The straight sinus is situated at the junction of the falx cerebri with the tentorium cerebelli. It is formed anteriorly by the confluence of the great cerebral vein (Galen) and the inferior sagittal sinus. It ends posteriorly by turning laterally to form one of the transverse sinuses, usually the left.

31 The superior sagittal sinus is situated at the attached border of the falx cerebri. It lies in a space between the two layers of the dura. Thus, superiorly, is the periosteal dura. Inferolaterally are the two layers of the meningeal dura, coming together, inferiorly, to form the falx. The sinus is lined with endothelium. It contains no muscle in its wall and is valveless.

B Applied answers

32 A lateral plain X-ray of skull is the most valuable to see the size and shape of the hypophyseal fossa. A tumour of the pituitary gland may cause erosion of the clinoid processes and be seen enlarging the bony fossa in all directions.

33 Thrombosis of the superior sagittal dural venous sinus will cause blockage in the normal flow of CSF,

as most of this fluid drains via the arachnoid granulations into this sinus. Reduced drainage with continuing production of CSF will cause the pressure to rise rapidly, and within a day or so that patient may become comatose.

34 The arcuate eminence of the petrous temporal bone is caused by the underlying superior semicircular canal of the inner ear.

35 The nerve of the pterygoid canal consists of fibres from both the greater petrosal and the deep petrosal nerves. The deep petrosal carries sympathetic postganglionic fibres, which lie around the internal carotid artery as a plexus, and enter the skull through the carotid canal. The greater petrosal nerve carries preganglionic parasympathetic fibres which synapse in the pterygopalatine ganglion before supplying the lacrimal gland via the zygomatic and lacrimal nerves. It is also secretory to the mucosal glands of the nasal cavity and palate. A most important component of the greater petrosal nerve is taste fibres from the palate which pass through the pterygopalatine ganglion on their way to the cell bodies in the geniculate ganglion. The consequences of damage to all these fibres in the pterygoid canal would result in dryness of the nose and palate, reduced tear formation and loss of taste from the palate.

36 The tegmen tympani of the adult is a thin sheet of compact bone separating the cranial cavity above from the tympanic cavity below. Because the middle ear and mastoid air cells are a fairly common site of infection, it is an important wall of resistance to brain abscess formation. In the young, the unossified petrosquamous suture more easily allows direct spread of infection from the middle ear to the cranial meninges.

37 The cribriform plate is so called due to its numerous sieve-like (L. *cribrum* = sieve) holes on each side of the crista galli. It is through these tiny holes that the olfactory nerve fibres pass to the upper epithelium of the nasal cavity and, hence, how one detects smells. Any fracture here will slice the fragile olfactory nerve fibres as they pierce the horizontal plate of the ethmoid bone. Because the olfactory nerve is not a true cranial nerve but is essentially a direct cerebral outgrowth, it is covered

with the three meninges. Consequently, if the meninges are torn, the CSF may leak down the nose. CSF can be distinguished from a normal runny nose by the increased glucose content of the discharge.

38 By using the anterior fontanelle, which is open at this age. The needle is introduced at the lateral edge of the diamond-shaped area between the membranous portions of the frontal and parietal bones. Advanced vertically to the skin, it will soon pierce the dura mater and enter the subdural layer where the haematoma is situated. In the young infant the presence of fontanelles makes burr-holes unnecessary.

39 Some of the veins on the face, especially in the region around the eyes and deep tissues of the cheek, drain via the ophthalmic veins and pterygoid venous plexus into the cavernous dural venous sinus. This lies each side of the pituitary gland and contains within its walls cranial nerves III, IV, V and VI on their way to the eye. Consequently, pus on the cheek, if pressed into the veins draining towards the cavernous sinus, may precipitate cavernous sinus thrombosis. This in turn will lead to pressure on the VIth nerve and a lateral rectus palsy may be the result.

40 The dural venous sinuses are best seen as the late stages of either a carotid or a vertebral arteriogram. About 4 seconds following its injection, the contrast medium may be seen in the major veins and venous sinuses. They may be viewed in the standard anteroposterior and lateral views, the latter being easier for the inexperienced viewer to comprehend.

C Information answers

41 Both the dartos muscle and the platysma are situated in the superficial fascia. The similarity ends there, however, because the dartos is innervated by the autonomic nervous system, and the platysma by voluntary nerves, viz. the cervical branch of the facial nerve.

42 The skin over the angle of the jaw is supplied by the greater auricular nerve (C2, 3) a branch of the cervical plexus. Its territory is marked posteriorly

by the anterior border of sternocleidomastoid. Behind this line the skin is supplied by the lesser occipital nerve (C2), also a branch of the cervical plexus. Anterior to the parotid, the skin of the face is supplied by the trigeminal nerve.

43 The skin of the 'moustache' area of the upper lip is supplied by the infraorbital branch of the trigeminal nerve. It is a branch of the nerve of the maxillary process in development. The central area of the upper lip, the philtrum, belongs to the frontonasal process which, in development, is supplied by the ophthalmic division of the trigeminal nerve. However, as development proceeds, this area is 'taken over' by the maxillary nerve. In cases of non-fusion of the frontonasal and maxillary processes (hare-lip), the philtrum retains its ancient nerve supply.

44 The muscles of facial expression are conveniently divided into sphincters and dilators. All are supplied by the facial nerve. The sphincters surround the eyelids and the mouth. They are termed orbicularis oculi and orbicularis oris, respectively.

45 The scalp consists of five layers. Most superficial is the skin. Below this is a layer of fibrofatty connective tissues, the septa of which unite the skin to the next deepest layer – the aponeurosis of the occipitofrontalis muscle (galea aponeurotica). Below the aponeurosis is a layer of loose areolar tissue which separates it from the pericranium (periosteum) of the skull. It is through this loose areolar layer that the layers of the scalp become separated in scalping. The layers of the scalp may be easily memorised, as the letters SCALP stand for each layer in turn:-

    S   skin
    C   connective tissue
    A   aponeurosis
    L   loose connective tissue
    P   periosteum.

46 The arterial supply of the scalp is plentiful and there are free anastomoses between the individual arteries. The arteries on each side (from anterior to posterior) are the supratrochlear, the supraorbital, the superficial temporal, the posterior auricular and the occipital. The supraorbital and

supratrochlear are branches of the ophthalmic artery that accompany the corresponding nerves. The ophthalmic artery arises from the internal carotid. The superficial temporal, posterior auricular and occipital arteries are all branches of the external carotid. The superficial temporal artery accompanies the auriculotemporal nerve.

47 The secretomotor supply to the parotid gland is via the parasympathetic system. The preganglionic pathway commences at the inferior salivary nucleus and leaves the brain in the glossopharyngeal nerve. The nerve fibres leave the glossopharyngeal nerve in its tympanic branch and pass through the floor of the tympanic cavity to the middle ear. Here they disperse to form the tympanic plexus but are reassembled as the lesser petrosal nerve. This nerve runs in a groove on the petrous temporal bone under the dura of the middle cranial fossa. It leaves the cranium via the foramen ovale to reach the otic ganglion immediately below. Here the parasympathetic secretomotor fibres synapse. The postganglionic fibres are distributed to the gland via the auriculotemporal branch of the mandibular nerve.

48 From superficial to deep, the structures which lie within the parotid gland are: the facial nerve, the retromandibular vein and the external carotid artery. The facial nerve leaves the skull via the stylomastoid foramen and enters the posterior part of the gland. As it runs forwards, within the gland, it breaks up into its five terminal branches. Deep to the facial nerve is the retromandibular vein. Its plane splits the gland into superficial and deep areas. It is formed by the union of the maxillary and superficial temporal veins, and runs superoinferiorly. As it leaves the inferior aspect of the gland, it divides into anterior and posterior divisions. Deep to the retromandibular vein is the external carotid artery. It runs inferosuperiorly. At the level of the neck of the mandible, it divides into its two terminal branches: the superficial temporal and the maxillary arteries.

C Applied answers

49 The orbicularis oculi muscle is the sphincter of the palpebral fissure. It consists of an inner

palpebral portion, within the lids, and an outer orbital portion, which comprises a complete ring and is attached only to the medial palpebral ligament. This attachment means that upon contraction of the muscle, as in blinking, tears are forced from the lateral aspect of the palpebral fissure, where they are secreted by the lacrimal gland, towards the medial aspect, where they drain via the lacrimal puncta. Damage causes the lower lid to fall away from the eye (ectropion). A stagnant pool of tears will then form in the lower fornix, which will eventually spill over the paralysed lower lid. Infection may occur in this pool, and the resulting conjunctivitis will increase the secretion of tears with further weeping. Paralysis of the orbicularis oculi will result in lack of protection of the cornea, which may dry out and subsequently ulcerate.

50 Herpes zoster tends to map out dermatomes. In this case we are dealing with the distribution from the infraorbital, zygomaticofacial and zygomaticotemporal nerves, these being the terminal sensory branches of the maxillary nerve. The area of distribution is developmentally the maxillary process, and consists of the skin of the cheek below the eye and above the upper lip, but not including the nose or philtrum because these develop from the frontonasal process and are therefore supplied by the ophthalmic division. The maxillary region also includes a small area superolateral to the eye, so the overall supply is a wing-shaped area. It is bounded above by branches of the ophthalmic and below by branches of the mandibular divisions of the trigeminal nerve.

51 The sphincters of the face (i.e. orbicularis oris and the two orbicularis oculi) are the most important facial muscles. Not only do they contribute to expression but they are also the anchors into which the majority of the other small facial muscles attach. Orbicularis oculi is the protector of the cornea and is therefore necessary for normal vision. Loss of orbicularis oris may lead to a mouth which drools uncontrollably, and is commonly seen in a facial nerve palsy (Bell).

52 The vestibule is the area between the cheek and teeth which is normally cleared of food by the

actions of the buccinator muscle and the tongue. A VIIth cranial nerve lesion will result in paralysis of buccinator and consequent collection of food within the vestibule. It will also lead to drooling from the corner of the mouth.

53 A lesion of V_1 will result in the cornea being insensitive to touch. Thus specks of dust or grit will not be felt in the eye. This may quickly lead to corneal ulceration. It is for this reason that after anaesthetising an eye it is most important to cover the patient's eye or he may collect more foreign bodies than he started with when coming to the surgery!

54 Collections of pus in the scalp tend to lie below the aponeurosis in the layer of loose areolar connective tissue but, due to the attachments of the occipitofrontalis, the abscess cannot track laterally. Fluid lying below the muscle layer can, however, track anteriorly to form an orbital swelling, e.g. blood causing a haematoma.

55 It is through the loose areolar tissue layer that a person is scalped.

56 A generalised scalp infection will drain to the horizontal group of lymph nodes, which are a ring of nodes lying along the superior limits of the cervical investing fascia. These are the submandibular, buccal, parotid, mastoid and occipital nodes. The majority of these groups of lymph nodes eventually drain into the chain of deep cervical nodes which are situated along the internal jugular vein.

57 The valveless emissary veins connect the intracranial venous sinuses with the superficial veins of the scalp. Therefore an abscess on the scalp may well pass infection into an intracranial sinus, thus forming a brain abscess.

58 Mumps is a viral inflammation of the parotid gland and this in itself is enough to cause pain. However, swelling of the parotid gland is made even more painful by the anatomical fact that the deep investing cervical (polo-neck) fascia splits to enclose the gland. Swelling therefore causes tension

within this restricted fascial plane. Movements of the inflamed fascia such as opening the jaw or yawning result in further pain.

59 The VIIth cranial nerve (facial) exits the skull through the stylomastoid foramen and turns anteriorly to pass right through the substance of the parotid gland, where it splits into its five main branches. It is here likened to a goose's foot (pes anserinus) and any tumour of the parotid gland may easily involve one or more branches which supply the muscles of facial expression.

60 It depends where the cut is performed, for it is along the auriculotemporal nerve that the postganglionic parasympathetic secretomotor fibres reach the parotid gland from the otic ganglion. If it is made high up on the temporal region, a lesion of the auriculotemporal nerve will only cause loss of scalp sensation; if it is made deep in the temporal fossa, proximal to its parotid branches, it will cause a loss of parotid secretions.

61 The opening of the parotid duct (Stenson) is in the vestibule above the second upper molar tooth. To palpate the duct, ask your patient to clench his teeth and the duct can be easily rolled on the contracted masseter muscle just a finger's breadth below and parallel to the zygomatic arch.

D Information answers

62 The middle meningeal artery is encircled by the two roots of the auriculotemporal nerve. This nerve is a branch of the posterior division of the mandibular part of V. It runs deep to the neck of the mandible and then turns upwards between the external auditory meatus and the temporomandibular joint. The nerve is sensory to the skin of the auricle, external auditory meatus, tympanic membrane and that skin overlying the parotid and temporal aspect of the scalp. It carries postganglionic parasympathetic fibres from the otic ganglion to the parotid gland.

63 The facial nerve gives off a branch within the petrous temporal bone at the level of the geniculate ganglion. This is the greater petrosal nerve, and it

contains preganglionic parasympathetic fibres destined for the pterygopalatine ganglion. It runs forwards in a groove in the petrous temporal bone, immediately deep to the dura of the middle cranial fossa, to reach the foramen lacerum. Here it is joined by the deep petrosal nerve. The deep petrosal nerve is a branch of the sympathetic plexus surrounding the internal carotid artery. The parasympathetic and sympathetic elements fuse to form the nerve of the pterygoid canal (Vidian nerve). This enters the pterygoid canal, which is a bony tunnel situated at the base of the pterygoid process of the sphenoid and running anteriorly through it to the pterygopalatine fossa. Here the nerve enters the pterygopalatine ganglion and the parasympathetic elements synapse. Postganglionic fibres are distributed to the lacrimal and nasal glands.

64 The middle meningeal artery arises from the first part of the maxillary artery. It runs upwards, superficial to the sphenomandibular ligament and deep to the lateral pterygoid muscle. It passes between the two roots of the auriculotemporal nerve to enter the skull via the foramen spinosum. Within the skull it divides into anterior and posterior divisions on the squamous temporal bone near the pterion, an important surface marking. The anterior branch crosses the greater wing of the sphenoid to lie on the parietal bone, where it runs adjacent to the precentral (motor) sulcus. The posterior branch runs onto the posterior aspect of the parietal bone. Both branches supply bone and meninges.

65 The lateral pterygoid arises by two heads. The upper head arises from the infratemporal surface of the greater wing of the sphenoid, the lower head from the lateral aspect of the lateral pterygoid plate. The fibres run backwards and converge to form a tendon, which is inserted into the anterior aspect of the neck of the mandible and into the articular disc of the temporomandibular joint. Contraction of this muscle pulls the combined mandible and disc forwards (protrusion of the jaw). The head of the mandible leaves the articular fossa and comes to lie on the articular eminence as this anterior movement proceeds. Protrusion of the jaw is a necessary accompaniment to opening the mouth (depression of the jaw). If it did not occur, the angle

of the jaw would press on the parotid gland and associated structures every time the mouth was opened.

66 The temporomandibular joint is a synovial joint of the condylar variety. The bones involved are the articular fossa and eminence of the temporal bone and the head of the mandible. A fibrocartilaginous disc is interposed between these bones and is intimately connected to the tendon of the lateral pterygoid muscle. The joint is unusual in that both articular surfaces are covered in fibrocartilage instead of the usual hyaline variety.

67 At the temporomandibular joint the mandible may be depressed or elevated, protruded or retracted. Depression is a simple hinge movement between the head of the mandible and the combined disc and articular fossa. It is brought about by the contraction of the digastrics, geniohyoids and mylohyoids. It is always associated with protrusion, in order to prevent the angle of the mandible pressing upon posterior relations. Elevation, the opposite of depression, is a powerful movement brought about by the temporales, masseters and medial pterygoids. Protrusion of the mandible is a movement whereby the head of the mandible plus the disc are pulled bodily forwards onto the articular eminence. It is brought about by the lateral pterygoids. The reverse action is retraction. This is performed by the posterior fibres of temporalis.

68 The pterygoid hamulus is a small hook of bone at the inferior aspect of the medial pterygoid plate. The pterygomandibular raphe attaches to it and separates the buccinator muscle anteriorly from the superior constrictor muscle of the pharynx posteriorly. The extreme upper part of the superior constrictor attaches to the hamulus. The tendon of the tensor palati winds round the hamulus from lateral to medial, to reach the palate. A bursa intervenes between the tendon and the bone.

69 The pterygomandibular raphe stretches between the hamulus of the medial pterygoid plate superiorly and the posterior end of the mylohyoid line of the mandible inferiorly. It is merely the fibrous junction of two muscles: the buccinator anteriorly and the superior constrictor posteriorly.

D Applied answers

70 The artery involved in extradural bleeding and haematomata formation is the middle meningeal artery. The pterion is a useful surface marking for it, as the artery enters the skull through the foramen spinosum in the sphenoid bone. The middle meningeal artery, a branch of the maxillary artery, is often encircled by the auriculotemporal nerve.

71 The pterygoid venous plexus, which lies deep within the temporal fossa surrounding the lateral pterygoid muscle, is connected with both the ophthalmic and the anterior facial veins. Thus the plexus drains into the cavernous sinus, and so infection within this venous plexus may cause direct increased ophthalmic venous pressure or may even lead to a cavernous sinus thrombosis with resultant paralysis of extraocular muscles.

72 The mastoid process is little developed at birth, as the air sacs are still rudimentary and consequently the facial nerve is very exposed as it exits the stylomastoid foramen. It can be palpated in the newborn and it is not surprising that, if trapped beneath forceps during delivery, a temporary facial palsy may result.

73 The parotid duct lies a finger's breadth below and parallel to the zygomatic arch; it can be rolled on a tensed masseter, particularly along its anterior border before the duct pierces the buccinator muscle.

74 To perform an inferior dental block, the dentist must place the anaesthetic agent around the inferior alveolar (dental) nerve. This is usually performed just above the lingula of the mandible, which protects the mandibular foramen through which the nerve enters the bone. Due to the proximity of the other major sensory branch of the mandibular division of V, the lingual nerve, it is also not uncommon to cause numbness of the tongue.

75 The pterygopalatine ganglion, which lies in the pterygopalatine fossa, is the largest peripheral ganglion of the parasympathetic system. If stimulated, it stimulates 'hay fever' causing streaming of the eyes, running of the nose, and mucous secre-

tions from the palate. The preganglionic fibres reach the ganglion by way of the facial nerve and the greater petrosal nerve. The postganglionic fibres are then distributed along the zygomaticotemporal, lacrimal, palatine and nasal nerves to the lacrimal gland and mucous glands of the palate and nose.

76 Lateral views of the temporomandibular joint with the mouth open and closed reveal the articular tubercle and its relationship to the condyloid process (head) of the mandible. The external auditory meatus is also well shown in both these views, with the tympanic plate lying at its inferior margin, protecting it from the mandibular fossa and joint itself.

77 Boring lectures cause yawning amongst the audience, and yawning is a cause of jaw dislocation. The lateral pterygoid muscles protrude and depress the jaw. If the other muscles of mastication are relaxed, the head of the mandible may 'click' over the articular tubercle and dislocate anteriorly to lie just alongside the zygomatic arch.

78 Trismus is a painful tonic spasm of the jaw muscles as seen in diseases such as tetanus (lockjaw). The muscles involved are the medial and lateral pterygoids, temporalis and, most importantly, the masseter muscle.

E Information answers

79 The orbit is a four-sided pyramidal-shaped cavity, with its base anteriorly and its apex posteriorly. The orbital margin is divided equally between three bones: superiorly, the frontal bone; anteromedially, the maxilla; and laterally, the zygoma. The roof is formed by the orbital plate of the frontal bone. The medial wall is made up, from anterior to posterior, of the frontal process of the maxilla, the lacrimal bone, the ethmoid and the body of the sphenoid. It is exceedingly thin. The floor is composed of the orbital plate of the maxilla. The lateral wall is made up of the zygomatic bone anteriorly and the greater wing of the sphenoid posteriorly.

80 The superior orbital fissure connects the middle cranial fossa with the orbit. It transmits the motor

nerves to the muscles of the eyeball (oculomotor nerve III, trochlear nerve IV, abducent nerve VI) and the three branches of the ophthalmic division of the sensory trigeminal nerve (V_1) – the lacrimal nerve, the frontal nerve and the nasociliary nerve. Sympathetic fibres also gain access to the orbit via the superior orbital fissure. The inferior orbital fissure connects the pterygopalatine fossa with the orbit. It transmits two branches of the maxillary division of the trigeminal nerve (V_2): the infraorbital nerve and the zygomatic nerve.

81 Tenon's capsule surrounds the eyeball from the optic nerve as far anteriorly as the corneoscleral junction, and separating it from the orbital fat gives it a smooth surface on which to move. The tendons of the six ocular muscles pierce this sheath; the sheath being reflected on them as a tubal covering. The inferior part of the sheath, which passes under the eye, is thickened. It suspends the eye as if it were a sailor in a hammock, and is termed the suspensory ligament of the eye. The suspensory ligament is attached to the lateral and medial walls of the orbit. Their attachments are common to those of the lateral and medial check ligaments, which are prolongations of the tubal coverings of the lateral and medial recti to the orbital walls.

82 The tendinous ring is the origin of the four rectus muscles of the eye. Each arises from its respective part of the ring. The lateral rectus arises by two heads which fuse as they approach the eyeball.

83 The cornea is the major refracting surface of the eyeball, the lens being used for fine adjustment. It is transparent and has no blood supply.

84 The four recti (medial, lateral, superior and inferior) arise from their respective sides of the tendinous ring, and diverging are inserted into the corresponding regions of the sclera some 5 mm behind the corneoscleral junction. The medial rectus muscle is a pure adductor and the lateral rectus a pure abductor. However, the muscles are arranged around the plane of the orbital cones, which are not parallel but diverge as they pass anteriorly, and not around the planes of movement of the eyeball. This means that the superior and inferior rectus muscles pass medial to the vertical

axis of the eyeball's rotation and act as adductors. Thus the action of the superior rectus is to elevate and adduct the eyeball, and that of the inferior rectus is to depress and adduct it.

The two oblique muscles counteract the adducting tendency of the superior and inferior recti. The superior oblique muscle arises from the body of the sphenoid superomedial to the tendinous ring. As it passes anteriorly it becomes tendinous. The tendon is slung from a pulley (or trochlea) on the anteromedial wall of the eyeball. Its action is to turn the eye downwards and outwards (depression and abduction). The inferior oblique arises from the anteromedial part of the orbital floor and passes posterolaterally, to be inserted into the lateral aspect of the eyeball. Its action is to elevate and abduct the eye so that it looks upwards and laterally.

85 The muscles of the iris consist of circular fibres and radial fibres. The circular fibres constrict the pupil. They are supplied by parasympathetic nerves which run in the oculomotor nerve, synapse in the ciliary ganglion and pass to the eye in the short ciliary nerves. The radial fibres dilate the pupil. They are supplied by sympathetic fibres which, having already synapsed in the superior cervical ganglion, pass to the internal carotid plexus and from there to the ciliary ganglion. Then they pass 'non-stop' through the ganglion to the eyeball, via the short ciliary nerves. Some sympathetic fibres pass from the nasociliary nerve to its long ciliary branches and thus to the eyeball. These latter sympathetic fibres reach the trigeminal nerve by passing to it from the internal carotid plexus in the cavernous sinus.

86 The optic nerve is accompanied along its entire course by sheaths of dura, arachnoid and pia. The meninges eventually fuse with the sclera of the eyeball. It is therefore surrounded by the subarachnoid space along its entire course. True cranial nerves possess no such prolongation of the subarachnoid space, the dura fusing with the perineurium soon after their origin. The optic nerve should therefore be regarded as a diverticulum of the brain and not a true cranial nerve.

87 The ciliary ganglion is situated in the posterior portion of the orbit, immediately lateral to the optic

nerve. It contains the synapses for parasympathetic nerves passing to the intrinsic muscles of the eyeball.

88 The ciliary ganglion is a parasympathetic ganglion. It receives preganglionic fibres from the oculomotor nerve via the branch to the inferior oblique muscle. In the ganglion the fibres synapse. Postganglionic fibres run to the eyeball via the 12 short ciliary nerves. They supply the ciliary muscle and the circular fibres of the iris (sphincter pupillae).

89 Both the long and short ciliary nerves contain sensory fibres. However, the long ciliary nerves also contain sympathetic fibres destined for the dilator pupillae. The short ciliary nerves contain both sympathetic and parasympathetic fibres for the dilator and sphincter pupillae, respectively.

E Applied answers

90 A stye is an infection of a sebaceous gland situated at the base of the eyelash. A Meibomian cyst is caused by a blocked tarsal gland which lies just posterior to the eyelash within the lids. These glands secrete an oily material which readily forms a noticeable cystic swelling if their drainage is impaired. These cysts often go unnoticed until infected, when they may easily be confused with a stye.

91 Ptosis may be congenital but there are two anatomical causes due to the two different muscles involved in keeping the upper eyelid raised. The greater portion of muscle fibres are the voluntary fibres of levator palpebrae superioris but the smooth muscle fibres of the superior tarsal muscle also contribute to the normal width of the palpebral fissure. The voluntary muscle is innervated by the IIIrd cranial nerve, whilst the smooth muscle is controlled by sympathetic fibres from the superior cervical ganglion. Damage to either of these nerves will cause ptosis.

92 Papilloedema is caused by an increase in CSF pressure. The reason why it can be readily seen at the optic disc is that the optic nerve is not a true cranial nerve but a brain extension and is therefore

surrounded by the three meninges. A rise in intracranial pressure will be transmitted therefore along the subarachnoid space and seen on the retina as swelling of the optic disc.

93 To test accommodation, your patient looks into the distance and is then asked to focus on a near object such as a pin. Three things happen: the eyes focus, the pupils constrict and the eyes converge. The pupillary and focusing changes are through afferent connections of the optic nerve with the pretectal area and efferent preganglionic parasympathetic neurons from the Edinger-Westphal nucleus. These efferent fibres, carried by the IIIrd cranial nerve, synapse in the ciliary ganglion and the postganglionic fibres pass to the ciliary and constrictor pupillae muscles.

94 Central retinal artery thrombosis causes instant blindness, as this is the only arterial supply to the retina and is a classic example of an 'end artery'. The central retinal artery is easily seen on ophthalmoscopy, as it leaves the optic nerve at the disc and divides into its temporal and nasal branches which supply the retina in a quandrantic manner.

95 The ptosis is likely to be due to sympathetic or IIIrd cranial nerve problems. The dilated pupil may be due to loss of parasympathetic control, and the position of the eyeball indicates a loss of those muscles supplied by the IIIrd cranial nerve (i.e. medial rectus, inferior oblique, superior and inferior recti). All these signs can be explained by damage to the oculomotor nerve (III) and its parasympathetic fibres.

96 The integrity of the sympathetic chain must be checked in cases of ptosis. An interruption in the sympathetic outflow from T1 may be caused by tumours in the apex of the lung. These tumours may also involve the brachial plexus and innervation of the hand (Pancoast's tumour).

97 The blind spot is where the fibres of the optic nerve pass through the retinal layers. It is the pale area seen on ophthalmoscopy as the optic disc and is of course also the point of entry of the retinal artery. It lies on the nasal side of the fundus.

98 The normal pressure of the aqueous humour is about 25 mmHg. A blockage in its normal drainage mechanism may cause acute glaucoma. Secreted by the ciliary body, the aqueous humour flows from posterior to anterior chambers before draining at the lateral angle of the iris into the corneoscleral junction, where lies an endothelial-lined canal (Schlemm). The fluid drains across the endothelial lining and into the ocular veins. This canal is so situated that prolonged dilation of the pupil may cause an impedance to the flow of aqueous humour in some people who possess a particularly shallow anterior chamber, due to bunching of the iris against its opening.

99 This is a VIth cranial nerve palsy affecting the lateral rectus muscle, its only motor supply. Due to the long intracranial length of this slender nerve, the condition of lateral rectus palsy may be caused by increased intracranial pressure from whatever origin.

F Information answers

100 The nasal septum consists of the perpendicular plate of the ethmoid superiorly, the vomer posteroinferiorly and the septal cartilage anteriorly.

101 The functions of the nasal cavity can be subdivided into those concerned with respiration and those concerned with olfaction. The respiratory mucous membrane covering the lower part of the nasal cavity has a rich blood supply. This serves to warm and moisten the inhaled air. The mucus secreted by this epithelium traps large particles such as dust, which is wafted to the pharynx by cilia and then swallowed. The olfactory mucous membrane lines the sphenoethmoidal recess and the upper part of the superior concha and adjacent septum. It contains the nerve endings of the olfactory nerves. For the detection of smell, the substances involved must be in solution. This area is therefore supplied with numerous serous glands. The paranasal air sinuses lighten the skull. They are supposed to impart a resonant quality to the voice in rather the same way as a soundbox does in a stringed instrument.

102 The lateral wall of the nose is divided up by three projections: the superior, middle and inferior

conchae. Above the superior concha is the sphenoethmoidal recess. Between the conchae are the superior and middle meati. Below the inferior concha is the inferior meatus. The sphenoidal air sinuses open into the sphenoethmoidal recess. The posterior cells of the ethmoidal sinus enter into the superior meatus. The middle meatus has a prominence on it – the bulla ethmoidalis. The middle ethmoidal cells open on its superior aspect. Below the bulla is a semicircular furrow – the hiatus semilunaris.

The anterior ethmoidal cells, and their continuation, the frontal sinuses, open into the anterior aspect of the hiatus. The maxillary sinus opens into the posterior aspect. Infection may easily spread therefore from frontal sinuses to maxillary sinuses via the hiatus semilunaris.

The inferior meatus receives the nasolacrimal duct, guarded by a valve of mucous membrane. This prevents air being blown in a retrograde manner up the duct.

103 Five arteries anastomose on the anteroinferior part of the nasal septum. This region is termed 'Little's area' and is a frequent source of epistaxis. The arteries are, first, the septal branch of the sphenopalatine artery, which is in turn derived from the maxillary. It enters the nose posteriorly from the pterygopalatine fossa via the sphenopalatine foramen. Second and third are the anterior ethmoidal and posterior ethmoidal arteries, both branches of the ophthalmic artery. They leave the orbit via foramina in its medial wall. Fourth is the great palatine artery, a branch of the maxillary. This artery reaches the nose via the incisive foramen in the palate. Last is the septal branch of the superior labial artery, which is in turn a branch of the facial artery.

104 The hard palate is supplied with sensory nerves from the maxillary division of the trigeminal via the pterygopalatine ganglion. These nerves are the greater palatine, the lesser palatine and the nasopalatine. The greater and lesser palatine nerves descend in the palatine canal and, passing through the palatine foramen, appear on the inferior surface of the back of the hard palate. The greater palatine nerves pass anteriorly. The lesser palatine nerves pass posteriorly and onto the anterior aspect of the soft palate. The nasopalatine nerves descend on the

vomer and pass through the incisive canal onto the anterior part of the hard palate.

The soft palate is supplied by the pharyngeal plexus. This is derived from the glossopharyngeal nerves, vagus nerves and sympathetic system. The glossopharyngeal nerve supplies the soft palate with sensory fibres. The muscles of the soft palate are supplied by the vagus (or more correctly, the cranial part of the accessory nerve that gives all of its fibres to the vagus), except tensor palati which is supplied by the trigeminal nerve (mandibular division).

105 The palatine tonsils are aggregates of lymphoid tissue situated at the posterolateral aspect of the oral cavity. Each tonsil lies between the palatoglossal arch anteriorly and the palatopharyngeal arch posteriorly. It lies on some loose areolar tissue which separates it from the superior constrictor of the pharynx. Superior to the tonsil is the soft palate; inferiorly, the posterior part of the tongue. The tonsil is largely supplied with blood by the tonsillar artery, a branch of the facial artery, which is situated immediately deep to the superior constrictor.

106 Waldeyer's ring comprises the nasopharyngeal adenoids, the palatine tonsils and the aggregates of lymphoid tissue on the dorsum of the tongue (lingual tonsil). Together they comprise a complete ring of lymphoid tissue surrounding the oronasal aperture of the body.

107 The pterygopalatine ganglion is situated in the pterygopalatine fossa and is suspended from the maxillary nerve. It is comprised of three roots: parasympathetic, sympathetic and sensory. Only the parasympathetic fibres synapse in the ganglion; the rest pass straight through.

The parasympathetic fibres originate in the facial nerve but leave it in the petrous temporal bone as the greater petrosal nerve. This unites at the foramen lacerum with the sympathetic deep petrosal nerve, to form the nerve of the pterygoid canal. Passing through the pterygoid canal, both sets of fibres reach the ganglion. Having synapsed, the parasympathetic fibres are distributed to the glands of the nose and pharynx and to the lacrimal gland via the zygomatic nerve. They are secretomotor to these glands.

The sympathetic root is comprised of the sympathetic fibres of the deep petrosal nerve. They are postganglionic, having already synapsed in the superior cervical ganglion.

Fibres from the maxillary nerve form the sensory root. They pass straight through the ganglion and form the bulk of its branches.

108 There are three constrictor muscles in the pharynx – superior, middle and inferior. The lowest fibres of the superior constrictor are overlapped by the upper fibres of the middle constrictor. Similarly, the upper fibres of the inferior constrictor overlap the lower fibres of the middle constrictor. The musculature of the pharynx therefore resembles three straight beer glasses (or flowerpots, for those horticulturally minded), placed one inside the other.

109 The inferior constrictor consists of two parts. The upper part arises from the oblique line on the thyroid cartilage. The fibres pass backwards to be inserted into the midline posterior pharyngeal raphe. It is sometimes termed the thyropharyngeus.

The lower part arises from the cricoid cartilage and forms a ring of muscle. It acts as a sphincter for the upper end of the oesophagus and, indeed, its fibres blend with those of the oesophagus. It is sometimes called cricopharyngeus. A potential weakness is present between these two portions of the inferior constrictor called Killian's dehiscence. It is here that pharyngeal diverticulae may develop.

110 The pharyngeal plexus is derived from branches of the glossopharyngeal nerve, the sympathetic system and the vagus nerve (strictly the accessory). It supplies all the muscles of the pharynx and soft palate except stylopharyngeus, which is supplied from the glossopharyngeal direct, and the tensor palati, which is supplied by the mandibular division of the trigeminal. The musculature of the tongue, both intrinsic and extrinsic, is supplied by the hypoglossal nerve, except for palatoglossus which is supplied by the pharyngeal plexus.

111 The glossopharyngeal nerve supplies one muscle only – the stylopharyngeus. The nerve descends in the neck in the carotid sheath. It then winds round the stylopharyngeus, supplying it as it does so.

112 The hyoid bone moves upwards and forwards during deglutition by the contraction of mylohyoid, geniohyoid, digastric and stylohyoid, thus compressing the food bolus between the tongue and the palate. Simultaneously, the wave of contraction spreads backwards from the anterior part of the tongue. As a result of these actions the food bolus is propelled towards the pharynx.

113 The epiglottis is connected to the posterior of the tongue, in the midline, by a fold of mucous membrane – the median glossoepiglottic fold. Two further mucosal folds connect the lateral aspects of the epiglottis with the pharyngeal wall – the lateral glossoepiglottic folds. Between these three folds are two depressions – the valleculae – which are therefore situated between the tongue and epiglottis, on either side of the midline.

F Applied answers

114 In the younger age group the majority of nose bleeds arise from the anterior septal region where numerous arteries anastomose. These are the anterior and posterior ethmoidal, sphenopalatine and greater palatine arteries as well as the superior labial branch of the facial artery. The bleeding is often due to playing with or picking the nasal septum and usually comes from this richly vascular area (Little's area, Kiesselbach's area). Bleeding more posteriorly from the territory of the greater palatine artery is often not controlled by simple pressure or nasal packing and may require ligation of the maxillary artery or even the external carotid artery.

115 The drainage of the frontal paranasal air sinus is by gravity to the middle meatus in the lateral wall of nose. Here, just below the bulla ethmoidalis in the hiatus semilunaris, opens the maxillary sinus. In the latter case, the ostium is near the roof of its cavity. It is therefore easy for infection in the frontal sinuses to drain straight down into the maxillary sinus, which has relatively poor drainage. The floor of the maxillary sinus (Highmore's antrum) lies adjacent to the roots and alveolar processes of the upper teeth. Pain from infection in the maxillary sinus is thus quite often indistinguishable from that of dental origin.

116 A technique for allowing a good view of the nasopharynx, posterior rhinoscopy is performed by passing a warmed angled mirror through the open mouth and fauces. Normally one would hope to see the posterior nasal septum (vomer) and the conchae in each nasal cavity as well as the tubal swellings guarding the auditory tubes (Eustachian, pharyngotympanic) superolaterally. In the roof, especially in the young child, one may see the pharyngeal tonsils or adenoids.

117 The pharyngotympanic or auditory tube equalises pressure in the middle ear, being opened each time one swallows. Its medial two-thirds is cartilaginous and externally, whilst its lateral portion is bony in the adult. In the adult, it has a downward path to the nasopharynx from the middle ear, thus allowing mucous secretions to drain freely. In children, however, not only is the tube relatively narrow but it also lies almost horizontal, making drainage difficult. Where it opens in the nasopharynx is the pharyngeal tonsil or adenoid, and any enlargement of this lymphoid mass will obstruct the drainage from the tube. Chronic obstruction may cause a collection of sticky mucus in the middle ear (glue ear), which is a ready medium for bacterial growth. In children, whose incidence of upper respiratory tract infection is already high, this may be a recurrent clinical problem.

118 A 'gag' reflex, often provoked by examination of the posterior nasopharynx, is due to stimulation of the glossopharyngeal nerve (IXth cranial). The sensory innervation of the fauces and movements of the pharynx are through the pharyngeal plexus, which has input from IX, X and the cranial portion of XI as well as a sympathetic contribution.

119 An enlarged pharyngeal tonsil or adenoid is a cause of auditory tube blockage, leading to recurrent ear infections. If particularly large, its removal may improve mucus drainage of the tympanic cavity and thus reduce the number of middle ear infections.

120 The jugulodigastric lymph node is normally palpable at the inferior angle of the jaw in acute tonsillitis.

121 One would suppose that removal of the epiglottis might lead to inhalation of food through an unprotected larynx. In fact this is not so because the purse-string-like action of the aryepiglottic folds is a much more efficient barrier and thus little inconvenience is felt after epiglottis removal, provided the aryepiglottic folds are repaired.

122 'Hypopharynx' is a clinical term used to describe the laryngopharynx. Malignant tumours of this region are mainly of epithelial origin, and two anatomical sites are particularly easy to miss. These are the valleculae at the root of the tongue and the piriform fossae which lie on either side of the laryngeal opening. Indirect laryngoscopy is performed through the open mouth and this simple procedure can check the hypopharynx and valleculae, as well as the vocal cords themselves. If the patient produces a high pitched 'ee-ee' the larynx is raised and the vocal cords are seen to adduct. When phonation ceases, the cords abduct to reveal the upper tracheal rings through the rima glottidis.

123 A pharyngeal pouch, or diverticulum, appears between the two parts of the inferior constrictor muscle. The inferior constrictor has a cricopharyngeal part and a thyropharyngeal portion, the latter fibres being oblique compared with the horizontal cricopharyngeal fibres. Posteriorly between these two is a weakness (Killian's dehiscence) through which a diverticulum or pouch may bulge. It is seen more commonly on the left and is thought to be due to incoordination of the cricopharyngeus muscle with resultant increased intrapharyngeal pressure.

124 The classic site for a fish bone to stick is in the piriform fossa or recess. These fossae are the portion of the laryngopharynx lying either side of the laryngeal inlet, and it is down these two channels that food passes, having been deflected from the larynx by the elevated epiglottis.

G Information answers

125 The styloid process, stylohyoid ligament and the upper half of the body of the hyoid bone are remnants of the cartilage of the second pharyngeal arch. The greater horn and the lower half of the

body of the hyoid bone are remnants of the third pharyngeal arch. Thus the hyoid bone is seen to be a hybrid, being derived from two embryological structures. The thyroid cartilage is the remnant of the cartilage of the fourth pharyngeal arch (and maybe the fifth as well). The cricoid and arytenoids are derived from the sixth arch cartilage. The stylomandibular ligament is merely a fascial condensation between the parotid and submandibular glands. Unlike the sphenomandibular and stylohyoid ligaments, it has no embryological significance.

126 The skeleton of the larynx consists of cartilages which are connected by membranes. The thyroid cartilage consists of two plates which meet anteriorly at a sharp angle. It is deficient posteriorly. Its posterior border has prominent cornua superiorly and inferiorly. The cricoid cartilage is signet-ring shaped. It articulates with the inferior cornu of thyroid cartilage. It is narrower anteriorly than posteriorly. The arytenoid cartilages are two small pyramids of cartilage which articulate on the posterosuperior aspect of the cricoid cartilage. They give attachment to the vocal cords and the muscles which move them. Unlike the other cartilages, the epiglottis is made up of elastic cartilage. It is leaf-shaped, being broader superiorly than inferiorly. Inferiorly it attaches to the posterior aspect of the front of the thyroid cartilage.

127 There are two sphincteric mechanisms of the larynx, designed to prevent food and fluids from entering the air passages. One mechanism is situated at the laryngeal inlet. A further back-up mechanism is present at the level of the cords. During swallowing, the larynx is pulled bodily upwards. The epiglottis is pushed backwards by the tongue and covers the laryngeal opening. The opening itself is narrowed by the contraction of the aryepiglottic muscles. Thus food passes over a covered and narrowed inlet. Should any food enter the larynx, however, the vocal cords may be adducted, preventing it from passing any further. Should it get past this second sphincter, the cough reflex is used. The cords are adducted and pressure builds up in the airways by muscular contraction. Sudden release of the cords produces an explosive rush of air which, hopefully, carries the offending particle with it to the exterior.

128 The posterior cricoarytenoid is the most vital muscle of the larynx. This is the only muscle capable of abducting the vocal cords and, hence, allowing air through the larynx. Its paralysis results in severe stridor. It runs from the posterior of the cricoid cartilage superolaterally to the muscular process of the arytenoid cartilage.

129 The sensory nerve supply to the larynx may be divided by the vocal cords into that part above the cords and that part below. The portion above the cords is supplied by the internal laryngeal branch of the superior laryngeal nerve, which is a branch of the vagus. Below the cords, sensation is provided by the recurrent laryngeal branch of the vagus. The recurrent laryngeal nerve supplies all the muscles of the larynx except for cricothyroid. This is supplied by the external laryngeal branch of the superior laryngeal nerve.

130 The two laminae of the thyroid cartilage meet anteriorly at the angle. This angle is prominent and is referred to as the laryngeal prominence (Adam's apple).

131 The levator glandulae thyroideae is a fibrous band which connects the isthmus of the thyroid gland to the hyoid bone. It may contain muscle.

132 Three arteries supply the thyroid gland: the superior thyroid artery, the inferior thyroid artery and the thyroidea ima artery. The thyroidea ima artery is not always present. The superior thyroid artery is a branch of the external carotid and runs vertically downwards to the upper pole of the gland in company with the external laryngeal nerve. The inferior thyroid artery is a branch of the thyrocervical trunk, which is in turn a branch of the first part of the subclavian artery. It ascends to the back of the gland and is usually tortuous. The recurrent laryngeal nerve is an intimate relation of this artery and may be anterior or posterior to it or even pass through its branches. The thyroidea ima is a branch of the aortic arch or the brachiocephalic artery. It ascends to the isthmus of the gland. The venous drainage of the thyroid gland is via the superior, middle and inferior thyroid veins. The superior and middle thyroid veins drain to the internal jugular vein. The inferior thyroid vein drains to the brachiocephalic veins.

133 The anterior triangle of the neck is bounded by the anterior border of the sternocleidomastoid laterally, the mandible superiorly and the midline of the neck anteriorly. The sternocleidomastoid, therefore, is the structure which divides the neck into its anterior and posterior triangles.

134 The deep fascia of the neck forms definite sheets, separating one area from another. The prevertebral and investing layers are two of these. The investing layer surrounds the entire neck and lies deep to the superficial fascia. It is attached below to the clavicle and manubrium, and above to the mandible, zygomatic arch, hyoid and base of the skull. It splits to enclose the parotid and submandibular glands and also the trapezius and sternocleidomastoid muscles. The prevertebral layer covers the muscle anterior to the cervical vertebral column, viz. longus cervicis and capitis. It also covers the scalenes and muscles of the back of the neck. Thus it forms the floor of the posterior triangle. Above, it is attached to the skull. Below, it fuses with the anterior longitudinal ligament at about T4. This limits the spread of pus into the inferior mediastinum.

135 Both the suprahyoid and infrahyoid muscles contract on swallowing, lifting the hyoid bone and larynx. Try it on yourself! The hyoid bone is raised during the first phase of deglutition by the suprahyoid muscles, forcing the food bolus to the back of the mouth. The larynx is lifted during the second phase by the infrahyoid muscles, preventing the influx of food into the airway.

136 Literally, 'ansa cervicalis' means the loop of the neck. The hypoglossal nerve contains not only hypoglossal fibres but also some fibres from C1. These leave the nerve in the carotid triangle as the descending branch (descendens hypoglossi). This descending branch joins with a descending branch of C2 and C3 to form a loop – the ansa cervicalis. The ansa supplies all the strap muscles of the neck except thyrohyoid, which receives C1 fibres that are given off separately from the XIIth cranial nerve.

137 The external carotid artery supplies the neck and the outside of the head with blood. The internal carotid artery has no branches outside the skull.

The branches of the external carotid artery in the neck are:
- Superior thyroid artery
- Ascending pharyngeal artery
- Lingual artery
- Facial artery
- Occipital artery
- Posterior auricular artery.

It ends at the level of the neck of the mandible by dividing into its two terminal branches – the maxillary artery and the superficial temporal artery.

G Applied answers

138 In both sexes during early childhood the larynx is similar in dimension. At puberty the male laryngeal cartilages enlarge considerably, whilst those of the female increase only very slightly. This enlargement accounts for the laryngeal prominence of the thyroid cartilage (Adam's apple) being easily seen and felt in the adult male. The deepening of the male adult voice is due to the increased cartilage size, especially in the anteroposterior diameter which is doubled at puberty.

139 The superior bony structure in the laryngeal skeleton is the hyoid. However, quite commonly in the elderly there is true ossification of the thyroid and cricoid cartilages which, with the arytenoids, are developed in hyaline cartilage. This is rarely seen in the young and does not affect the corniculate, cuneiform or the apex of the arytenoid cartilages, which all develop in elastic fibrocartilage and cannot therefore ossify.

140 The hyoid lies at the level of C2–3 vertebral bodies. The isthmus of the thyroid gland lies anterior to second and third tracheal rings (C7–T1) where it is separated from the strap muscles by the pretracheal fascia. The C6 vertebral level is where the larynx becomes the trachea and the pharynx continues as the oesophagus, and is the level of the cricoid cartilage.

141 The structures seen on indirect laryngoscopy are, first, the base of the tongue and lingual surface of the epiglottis and then, by instructing the patient to produce a high-pitched 'ee-ee', the vocal cords

are brought into view. On ceasing phonation the cords abduct, revealing the upper tracheal rings through the opened rima glottidis. The vestibular folds above the true cords are pink, fleshy folds of mucous membrane compared with the normal pearly white of the vocal cords. Before withdrawing the mirror, the piriform fossae and valleculae are checked for abnormalities.

142 The external laryngeal nerve supplies the cricothyroid muscle, paralysis of which causes weakness in the voice due to reduced vocal cord tension. It does not normally affect breathing. The recurrent laryngeal nerves supply all the intrinsic laryngeal muscles except for the cricothyroid and thus a bilateral paralysis results in the cords abducting, leaving only a 'glottic chink' through which inadequate air can pass. The patient has only a hoarse whisper and will suffer respiratory distress, possibly requiring a tracheostomy.

143 This is because the radiologist has asked the patient to say 'ee-ee' to stretch and adduct the cords. On relaxation and in normal quiet respiration the cords should abduct. If a cord is paralysed, abduction will not take place.

144 These procedures are performed in order to relieve acute airways obstruction above or at the level of the vocal cords. Laryngotomy is an opening made just above the cricoid cartilage through the cricothyroid ligament and is usually of a very temporary nature. Tracheostomy is a more permanent procedure, often performed by the surgeon, involving an opening into the trachea usually at the level of the second to fourth tracheal rings. In laryngotomy, besides the skin and an occasional anterior jugular vein, the only structure protecting the airway is the cricothyroid ligament itself. In the normal tracheostomy, however, one needs not only to retract the strap muscles of the neck but also to divide the very vascular isthmus of the thyroid. In children this procedure may endanger the left brachiocephalic vein or even the thymus, which often extends above the suprasternal notch. The secret to both procedures is to keep right in the middle.

145 The tortuous course of the inferior thyroid artery is readily seen on an arch aortogram and is due to

the fact that on every swallow the thyroid gland ascends a few centimetres and must naturally drag its arterial supply with it. If this artery had no capability to elongate, it would be traumatised.

146 Part of the pretracheal fascia forms the sheath of the thyroid gland and can be seen lying between the two sternohyoid muscles when the investing layer of deep cervical fascia has been incised. Superiorly, it is attached to the oblique line of the thyroid cartilage and the arch of the cricoid. Having enclosed the gland, it passes inferiorly to enclose the inferior thyroid veins and then blends with the posterior surface of the pericardium at the bifurcation of the trachea. Thus, when the laryngeal skeleton ascends during swallowing, so must the thyroid gland.

147 The very close but variable relationship between the inferior thyroid artery and the recurrent laryngeal nerves in the posterior part of the gland makes thyroid surgery, especially a thyroidectomy, a potential risk to normal speech. Prior to operation it is common practice to check that both cords are in working order so that if there is any problem postoperatively, one knows at least the origin of the lesion! If there is any preoperative abnormality then the surgeon is especially careful to leave the posterior region of the gland intact.

148 Between the first and second pharyngeal pouches the thyroid diverticulum originates during the fourth week of development. This diverticulum elongates caudally, forming the gland and leaving a solid stalk whose site of attachment can be seen as the foramen caecum of the tongue. An epithelial tract may remain from the dorsum of the tongue to the definitive gland (thyroglossal duct) and occasional cystic swellings may occur (thyroglossal cysts). If such a swelling is in the midline and elevates on protruding the tongue, then a thyroglossal cyst is the most likely diagnosis. These structures may be anywhere between the foramen caecum and the thyroid gland itself.

149 The two pairs of parathyroid glands develop from the third and fourth pharyngeal pouches. The glands from the third pouch usually migrate caudally with the thymus gland, overtaking the glands from the fourth pouch and coming to lie in

the fascial sheath in the upper part of the thyroid gland. Variations in number, size and position are common and aberrant third pouch parathyroid glands may descend with the thymus to a retrosternal position, instead of the more common site in the inferior lobe of the thyroid gland.

150 The carotid sheath is a fascial tube surrounding the common and internal carotid arteries and internal jugular vein. Lying within it posteriorly is the vagus nerve. The sheath is in close connection with the prevertebral fascia, with the sympathetic chain intervening. The surface markings are a thick band some 1.5–2 cm wide from the sternoclavicular joint to the tragus. The division of the common carotid lies 1 cm below and behind the greater cornu of the hyoid at C3–4 vertebral level.

151 The internal carotid artery has no branches in the neck, its first branch being the ophthalmic artery. The external carotid artery has numerous large cervical branches to the thyroid, tongue, pharynx, ear, occiput and face.

152 The vagus nerve, with its many branches to the larynx and pharynx, is an immediate posterior relation of the carotid artery, as is the sympathetic chain which lies just outside the carotid sheath on the prevertebral fascia. Both of these nerves may be damaged if the needle penetrates too deeply.

H Information answers

153 The vestibule of the mouth is that part which lies external to the gums and teeth. Anterior to it the lips are situated. Laterally is the cheek containing the buccinator. Superiorly and inferiorly the vestibule is limited by the reflection of the mucous membrane of the lips and cheek onto the gums.

154 The tongue is divided into an anterior two-thirds and a posterior one-third by the V-shaped sulcus terminalis. As regards general sensation, the anterior two-thirds is supplied by the lingual nerve (a branch of the mandibular division of the trigeminal); the posterior third is supplied by the glossopharyngeal nerve.

155 Taste fibres from the anterior two-thirds of the tongue travel in the lingual nerve to the infratemporal fossa. Here they leave it in the chorda tympani to join the facial nerve in the petrous temporal bone. Taste fibres from the posterior third run in the lingual branch of the glossopharyngeal nerve. This nerve also supplies the circumvallate papillae, which are situated immediately anterior to the sulcus terminalis. Taste fibres from the region of the epiglottis and valleculae run in the internal laryngeal branch of the vagus nerve.

156 The muscles of the tongue are divided into two groups: intrinsic and extrinsic. The intrinsic muscles are confined to the tongue itself, are not attached to any bone and serve to alter the shape of the organ. They are arranged in longitudinal, vertical and transverse bundles. The extrinsic muscles are attached to the surrounding skeleton and serve to move the tongue as a whole. They are the genioglossus and the paired hyoglossus, styloglossus and palatoglossus. The genioglossus runs from the genial tubercles, on the internal aspect of the mandible, to the tongue. It protrudes the tongue. The hyoglossus runs from the upper border of the hyoid, as a quadrilateral sheet, to the tongue. If the hyoid is fixed, contraction of this muscle depresses the tongue. The styloglossus runs from the tip of the styloid process to the tongue, pulling the tongue backwards and upwards upon contraction. The palatoglossus runs from the aponeurosis of the soft palate, down the lateral pharyngeal wall to the tongue. Together they act as a sphincter, closing the oropharyngeal orifice. All the muscles of the tongue (intrinsic and extrinsic) are supplied by the hypoglossal nerve except for palatoglossus, which is supplied by the pharyngeal plexus.

157 The thyroid gland develops as a diverticulum of the floor of the embryological pharynx. The developing tongue overgrows this diverticulum and hence it appears that the diverticulum arises from the tongue. It is therefore called the thyroglossal duct. As the thyroid develops it descends, and the thyroglossal duct atrophies. In the normal individual all that remains is a small pit on the apex of the 'V' of the sulcus terminalis. This is the foramen caecum. Sometimes, however, the thyroid

fails to descend and remains in the region of the foramen caecum. This is called a lingual thyroid. Sometimes the thyroglossal duct fails to atrophy completely. Cysts may form in those parts remaining.

158 The chorda tympani is a branch of the facial nerve. It leaves the facial nerve in the petrous temporal bone, crosses the tympanic cavity and leaves the skull via the petrotympanic fissure, to enter the infratemporal fossa. In the infratemporal fossa it joins the lingual nerve. The chorda tympani contains two kinds of fibres: taste fibres and preganglionic parasympathetic secretomotor fibres. The taste fibres originate from the anterior two-thirds of the tongue. Their cell bodies are situated in the geniculate ganglion of the facial nerve. The preganglionic parasympathetic fibres are destined for the submandibular ganglion. Here they synapse and are distributed to the sublingual and submandibular salivary glands.

159 The deciduous teeth do not start to erupt until the sixth month. Therefore a baby of 4 months would have no visible teeth at all. The central deciduous incisors erupt at about 6 months, the lateral incisors at 9 months. The first deciduous molars arrive at 12 months and the canines at 18 months. A child of 20 months would be expected, therefore, to possess all the deciduous incisors and canines and the first molars (the second deciduous molars do not usually erupt until 24 months of age). By the age of 12 years all the deciduous teeth have been replaced by the permanent dentition. Furthermore, all the permanent teeth have erupted by this age, except the third molars (wisdom teeth), which may not arrive until the thirtieth year. Thus a child of 12 years would have – in each half jaw – two incisors, one canine, two premolars and two molars. All of these teeth would be permanent.

160 All three of the major salivary glands (parotid, submandibular and sublingual) are in direct contact with the mandible. The parotid gland overlies the posterior border of the ramus of the mandible and its adjacent medial and lateral surfaces. The superficial part of the submandibular gland lies in immediate relationship to the submandibular fossa of the mandible. This is situated on the medial

aspect of the body, below the mylohyoid line. The sublingual gland lies in the sublingual fossa on the inner and posterior aspect of the mandible above the mylohyoid line. The following nerves, all directly related to the mandible, are branches of the mandibular nerve: the lingual nerve, the nerve to mylohyoid, the inferior alveolar nerve and the auriculotemporal nerve. The lingual nerve runs in a groove on the mandible for a short distance, related to the third molar tooth above. The inferior alveolar nerve enters the mandibular foramen and supplies the mandible, teeth and gums. It emerges at the mental foramen as the mental nerve. The nerve to mylohyoid, a branch of the inferior alveolar, runs on the medial aspect of the body of the mandible immediately below the mylohyoid line. It supplies the muscle from its superficial aspect. The auriculotemporal nerve runs posteriorly round the neck of the mandible to reach the temporal region.

161 The submandibular gland consists of two parts: a larger superficial part and a smaller deep part. These two parts are continuous around the free posterior border of mylohyoid, and hence the gland looks like a letter 'C' when removed. The submandibular duct leaves the gland at the anterior extremity of the deep part and passes anteriorly along the floor of the mouth lateral to the tongue, to open into the mouth at the summit of a papilla, situated immediately lateral to the frenulum of the tongue.

162 Preganglionic parasympathetic fibres from the superior salivary nucleus leave the brain in the nervus intermedius in the groove between the pons and the medulla oblongata. The nervus intermedius runs into the internal auditory meatus with nerves VII and VIII and fuses with the facial (VII). Within the petrous temporal bone the parasympathetic fibres destined for the submandibular gland leave the facial nerve in the chorda tympani, which crosses the tympanic membrane and leaves the skull at the petrotympanic fissure, to enter the infratemporal fossa. Here the chorda tympani joins the lingual nerve. Passing in the lingual nerve, the parasympathetic fibres reach the submandibular ganglion, which lies under the lingual nerve superficial to hyoglossus. In the ganglion the fibres synapse. The postganglionic fibres to the sub-

mandibular and sublingual glands run either directly or in branches of the lingual nerve.

163   In the submandibular region, on the hyoglossus muscle, the submandibular duct (Wharton) passes forward to enter the mouth. The lingual nerve crosses it from superior to inferior on its lateral aspect. It then immediately winds underneath the duct and crosses it from inferior to superior on its medial aspect, to reach the tongue. This 'double-crossing' of Wharton's duct by the lingual nerve is remembered by most students, as it is immortalised in a ribald rhyming mnemonic.

H Applied answers

164   A midline congenital tumour in the region of the foramen caecum, though rare, is quite likely to be a lingual thyroid remnant. This may turn out to be the only thyroid tissue the child possesses. The histology will consist of colloid-filled thyroid follicles whose walls are made up of a single layer of cuboid epithelium.

165   The genioglossus muscles originate from the genial tubercle, or mental spine, of the mandible and spread out like a fan backwards into the bulk of the tongue. In the unconscious patient the tongue falls back, blocking the airway, and the simplest way to stop this is to push the angle of the jaw forwards. In doing so, the genioglossi are pulled anteriorly and the bulk of the tongue is removed from the oropharynx.

166   Within the distribution of the trigeminal nerve, referred pain from one region to another is quite a common symptom. In this case, the lingual nerve to the tongue is being stimulated by the mouth lesion but the pain is being felt through the auriculotemporal nerve to the ear.

167   The anterior two-thirds of the tongue is drained by lymphatics to its ipsilateral submental and submandibular nodes, whereas the posterior third of the tongue has many lymphatic anastomoses across the midline, which drain to the upper deep cervical nodes. A posterior tumour is therefore more likely to spread to contralateral nodes. The rationale to the lymphatic drainage is due to a

partial midline septum in the anterior two-thirds.

168 When asked to put out her tongue, a right hypoglossal nerve palsy would be seen as an atrophy of the right side accompanied by deviation of the tongue towards the affected side.

169 In removing the lower wisdom teeth it is quite easy to traumatise the lingual nerve as it runs on the mylohyoid muscle just medial to the roots of the lower wisdom teeth. Damage to this nerve affects not only the sensory supply from the anterior two-thirds of the tongue but also the fibres carried by the chorda tympani. Proximal to the molar teeth, deep in the infratemporal fossa, the chorda tympani joins the lingual nerve carrying the special sensory (taste) fibres from the anterior part of the tongue, as well as the preganglionic parasympathetic secretomotor fibres to the submandibular and sublingual glands. Damage to these fibres will alter the normal salivary secretions, which patients may notice as a funny metallic taste. This taste is probably due to an imbalance of ions due to changed salivary concentrations.

170 The taste fibres in the chorda tympani are the peripheral processes of the geniculate ganglion cells and any tumour within the middle ear may affect either the chorda tympani or the ganglion itself. The geniculate ganglion lies above the promontory on the medial wall of the middle ear, whilst the chorda tympani lies between the malleus and incus on the medial side of the tympanic membrane.

171 In general, girls' teeth erupt slightly earlier than boys' and the lower jaw before the upper, but by the age of 22 the majority of students would have their third molar or wisdom teeth. If the third molars did not require removal due to impaction, a full set of adult teeth is 32 teeth. However, some people's wisdom teeth do not erupt until 24 years or beyond. The normal range for the appearance of these four teeth is 16–30 years.

172 The mental nerve, which supplies the lower lip, is a branch of the inferior alveolar nerve which lies inside the mandible supplying the teeth. Any jaw

fracture might therefore cause loss of sensation over the chin and lower lip.

173 The lower lip musculature is supplied by the cervical and mandibular branches of the facial nerve. In operating on a submandibular growth it may be impossible to avoid cutting through either or both of these nerves, leaving the patient with a drooped lip.

174 The radiologist only has to ask the patient to stick out his tongue. On putting the tip of the tongue up to the roof of the mouth, a prominent midline fold of mucous membrane – the frenulum – can be seen. On either side of the frenulum are the sublingual papillae, on the surface of which can be found the opening of the submandibular duct (Wharton) where the cannula can be inserted by the radiologist. Lying lateral to the papillae are the sublingual folds, beneath which lie the sublingual glands whose numerous ducts open along the crest of the folds. A few drops of lemon juice on the tongue may produce a jet of saliva from the submandibular duct orifices which lie just lateral to the frenulum.

I Information answers

175 The floor of the posterior triangle is formed by muscles, covered by the prevertebral layer of deep cervical fascia. From superior to inferior, these muscles are semispinalis capitis, splenius capitis, levator scapulae and the three scalenes (medius, posterior and anterior).

176 The posterior triangle of the neck is situated with its apex superiorly and its base inferiorly. It is bounded inferiorly by the middle third of the clavicle, anteriorly by the posterior border of the sternocleidomastoid and posteriorly by the anterior border of trapezius. Its roof is the investing layer of the deep cervical fascia, covered in turn by superficial fascia containing the platysma, and skin.

177 The deep cervical chain of nodes, which drain the structures of the head and neck, are situated under cover of the sternocleidomastoid muscle along the internal jugular vein. Clinically, they are difficult to feel unless the muscle is relaxed.

178 The sensory branches of the cervical plexus are the lesser occipital nerve, the greater auricular nerve, the transverse cervical nerve and the supraclavicular nerves. They all appear in the posterior triangle behind the posterior border of the sternocleidomastoid. The lesser occipital nerve (C2) hooks round the anterior aspect of the accessory nerve in the posterior triangle. It ascends up the posterior border of the sternocleidomastoid to supply skin behind the ear and around the mastoid process. The greater auricular nerve (C2, 3) appears in the posterior triangle at the posterior border of the sternocleidomastoid. It passes anterosuperiorly onto the muscle and ascends in company with the external jugular vein. It supplies skin over the angle of the mandible and the parotid region. The transverse cervical nerve (C2, 3) appears behind the posterior border of the sternocleidomastoid and passes anteriorly over the muscle to supply the skin of the front of the neck. The supraclavicular nerves (C3, 4) are usually three in number – lateral, intermediate and medial. They pass posteroinferiorly across the clavicle to supply the skin of the upper chest and deltoid region. They may actually pierce the clavicle.

179 C1 has no cutaneous branch and therefore normally has no dermatome, although it does supply sensation to the dura around the sigmoid sinus.

I Applied answers

180 The posterior triangle of the neck is bounded by the clavicle, sternocleidomastoid and trapezius muscles. Its superior half contains few structures of significance except for the spinal part of the accessory nerve and sensory branches of the cervical plexus. The inferior portion, however, contains not only the trunks of the brachial plexus, lying deep to omohyoid muscle, but also the subclavian vein and artery. Both of these vessels have numerous branches in the posterior triangle, the suprascapular and transverse cervical being two of the more vulnerable vessels.

181 The spinal part of the accessory nerve supplies the sternocleidomastoid and trapezius muscles. It traverses the posterior triangle lying rather

superficially in the cervical investing fascia and is quite easily damaged if an incision is made transversely across the triangle. Damage to the trapezius innervation makes elevation movements of the shoulder in abduction very difficult.

182 Leprosy is a systemic infection with a predilection for nerves, causing loss of sensation, depigmentation and thickening of peripheral nerves. These swollen nerves are felt as lumps or cords under the skin. An often visible abnormality is swelling of the greater auricular nerve (C2, 3) which supplies the angle of the jaw, mastoid region, side of the neck and both surfaces of the ear.

183 There are numerous inconstant venous channels between the brain and scalp but there are also at least six paired emissary veins, the most constant being those found in the parietal, mastoid and condylar regions. There is also often a large unpaired occipital emissary vein near the external occipital protuberance. There are, therefore, numerous venous connections of the occipital region with the intracranial venous sinuses, providing an easy path for the spread of infection.

184 Deep in the floor of the suboccipital triangle lies an important source of blood supply to the brain – the vertebral artery. Bounded by the small rotator muscles of the neck and lying on the posterior arch of the atlas, the vertebral artery ascends through the transverse foramen of C1, turns medially prior to penetrating the posterior atlanto-occipital membrane and then ascends through the foramen magnum.

J Information answers

185 The parietal pleura rises up into the neck to the level of the neck of the first rib. In this area it is termed the cervical pleura. Its surface marking is a curved line joining the sternoclavicular joint to the lateral extremity of the medial third of the clavicle. This line should extend 2.5 cm above the clavicle at its maximum convexity. The cervical pleura is covered by the suprapleural membrane (Sibson's fascia), which is attached to the inner margin of the first rib and separates the root of the

neck from the thorax. The subclavian vessels arch over this membrane.

186 The branches of the subclavian artery are: the vertebral artery, internal thoracic artery, thyrocervical trunk, costocervical trunk and dorsal scapular artery. With the exception of the dorsal scapular branch, on both sides, and the costocervical branch on the right, all the branches arise from the first part of the artery – i.e. medial to scalenus anterior. The vertebral artery enters the transverse foramen of C6 and ascends through successive foramina. It winds posterior to the lateral mass of the atlas and enters the foramen magnum. It supplies the posterior parts of the brain. The internal thoracic artery descends behind the costal cartilages of the ribs, supplying the intercostal spaces and surrounding structures. The thyrocervical trunk divides into inferior thyroid, suprascapular and superficial cervical arteries. The costocervical trunk divides into deep cervical and superior intercostal arteries. The dorsal scapular artery supplies a small amount of blood to the anastomosis around the scapula.

187 The vertebral artery arises from the first part of the subclavian artery. It runs upwards between the longus cervicis and scalene muscles, passes anterior to the transverse process of C7 and enters the transverse foramen of C6. It ascends in successive transverse foramina and on arising through that of C1, the artery winds behind its lateral mass, in the suboccipital triangle, then turns medially and pierces the posterior atlanto-occipital membrane to lie in the vertebral canal. It then passes upwards and enters the skull through the foramen magnum, then runs almost horizontally as it runs upwards, forwards and medially between the medulla and clivus. At the upper border of the medulla it fuses with its fellow to form the basilar artery. By convention, the vertebral artery is divided into four parts. The first part extends from the subclavian artery to the transverse process of C6. The second part is that where the artery lies in the transverse foramina. The third part is in the suboccipital triangle, and the fourth part is the intracranial portion.

188 The recurrent laryngeal nerves arise from the vagus nerves but at different levels on the two sides. The right recurrent laryngeal nerve arises from the right

vagus as it crosses anterior to the first part of the subclavian artery. It hooks round under the right subclavian artery and ascends between the trachea and oesophagus to the larynx. The left recurrent laryngeal nerve arises from the left vagus as the latter crosses anterior to the aortic arch. It loops beneath the ligamentum arteriosum and ascends between the trachea and oesophagus to the larynx. Developmentally, the recurrent laryngeal nerves are the nerves of the sixth pharyngeal pouches. They are, therefore, associated with the sixth pharyngeal arch arteries. As the neck enlarges and the heart descends, these vessels are pulled inferiorly and the nerves with them so the nerves now take a recurrent course downwards, round the vessel and upwards again. On the left side, the sixth arch artery becomes the ductus arteriosus. The left recurrent laryngeal nerve, therefore, runs round this vessel which, in adult life, is obliterated to a fibrous cord, the ligamentum arteriosum. On the right, the sixth and fifth arch arteries disappear. The right recurrent laryngeal nerve therefore loops round the next most superior vessel, which is the right fourth arch artery. This vessel develops into the proximal portion of the right subclavian artery, which the right recurrent laryngeal nerve is seen to run around.

189 The thoracic duct is a major lymphatic vessel recognised by its beaded appearance due to the presence of numerous valves. It commences at the upper border of the cisterna chyli, which is situated on the posterior abdominal wall at L1 on the right of the aorta. The thoracic duct ascends through the aortic hiatus of the diaphragm into the thorax. At the level of T4 it crosses to the left-hand side of the vertebral column, passing behind all the longitudinally running thoracic structures as it does so, and comes eventually to lie on the left of the oesophagus. In this position it continues to ascend into the neck. At the level of C7 it arches laterally, and then turns downwards, passing anterior to all the major arterial vessels of the neck, to end by draining into the confluence of the internal jugular and subclavian veins.

190 The cervical sympathetic trunk possesses three ganglia, termed the superior, middle and inferior cervical. The superior cervical ganglion is derived from cord segments T1–T4, the middle from

T5–T6 and the inferior from T7–T8. The inferior ganglion is frequently fused with the upper thoracic ganglion to form the stellate ganglion. The superior ganglion is situated at the level of the hyoid bone, the inferior around the neck of the first rib. The efferent branches are of three kinds: grey rami communicantes, direct branches and vascular branches. Each cervical spinal nerve receives a grey ramus communicans from its respective ganglion. The superior cervical ganglion sends direct branches to the first four cervical nerves, the pharynx and the cardiac plexus. It also sends vascular branches with the external carotid artery and its branches. A further branch, the internal carotid nerve, accompanies the internal carotid artery through the carotid canal. The middle cervical ganglion sends a direct branch to the thyroid gland along the inferior thyroid artery. The inferior ganglion also sends a direct branch to the cardiac plexus and vascular branches to the subclavian and vertebral arteries. The trunk itself divides on either side of the subclavian artery to form the ansa subclavia.

191 The scalenus anterior is a cervical muscle of the prevertebral group. This means that it lies posterior to any viscera or vessels in the neck. It is covered anteriorly by the prevertebral fascia. Between the muscle and the fascia is situated the phrenic nerve as it descends towards the thorax. Anterior to this musculoskeletal plane there is, in the neck, an arterial plane in which the transverse cervical artery is situated. The thoracic duct is anterior to this arterial plane. Thus the order to structures from posterior to anterior is: scalenus anterior, phrenic nerve, transverse cervical artery and thoracic duct.

192 The trunks of the brachial plexus and subclavian artery and vein all cross the superior surface of the first rib to enter the arm. The lower trunk of the brachial plexus is derived from roots C8 and T1, which arise on either side of the first rib. This means that the lower trunk is actually in contact with the rib's upper surface. The subclavian vessels also are in contact with the superior surface as they arch over the rib and descend into the arm. Of these structures, the subclavian vein is the most anterior. The subclavian artery is posterior to the vein but separated from it by the insertion of scalenus anterior at the scalene tubercle. The nerves are

posterior to these vessels. Thus the lower trunk of the brachial plexus lies posterior to the artery. In summary, the order of structures from anterior to posterior is: subclavian vein, subclavian artery and lower trunk of brachial plexus.

193   The first rib is the shortest of all the ribs and has the most pronounced curve. It is flattened superoinferiorly and hence possesses medial and lateral edges. Normal ribs are flattened medio-laterally and have superior and inferior edges. The first rib has a scalene tubercle on its medial border (edge) for the insertion of scalenus anterior. This muscle separates the subclavian artery posteriorly from the subclavian vein anteriorly at this point. Both the above vessels can be seen to groove the upper surface of the first rib, a feature possessed by no other rib. The head of the first rib is small and round. It articulates with one vertebral body only (that of T1) and hence bears only one articular facet.

J   Applied answers

194   Insertion of a needle into the angle formed by the confluence of the internal jugular and subclavian veins might well encounter the terminal portion of the thoracic duct. The brachiocephalic route for a central venous catheterisation on the left-hand side has been known to cause a chylothorax.

195   The muscles involved in shoulder abduction are primarily the deltoid with assistance from supraspinatus. The nerve supply to the deltoid is the axillary nerve, from the posterior cord of the brachial plexus, which winds around the neck of the humerus. The supraspinatus muscle, however, is innervated by the suprascapular nerve (C5, 6) which is derived directly from the upper trunk. Without the supraspinatus the initial component of abduction is weakened, and the more one tries to abduct the arm, the more one shrugs the shoulder. This is because the supraspinatus holds the humeral head into the glenoid fossa and stops it sliding in abduction.

196   The immediate posterior relationship of the great veins of the neck is the cupola of the pleura and apex of the lung. Posteriorly, this is the level of the

first rib. However, as the ribs slope downwards, the pleura extends above the first rib anteriorly. Just posterior to the sternocleidomastoid muscle the pleural dome lies some 2–3 cm above the medial third of the clavicle. In this region, the cupola of the pleura is strengthened by a fascial thickening called the suprapleural membrane (Sibson's fascia). Any catheterisation that misses the veins and continues posteriorly may produce a pneumothorax.

197 The inferior cervical and first thoracic sympathetic ganglia are often amalgamated to form the cervicothoracic or stellate ganglion just posterior to the origin of the vertebral artery. A cervical or upper thoracic sympathectomy will cut off the sympathetic innervation to the whole of the head and neck of the same side. This leaves the patient with a hot, dry, flushed face with no capacity to sweat (anhydrosis), ptosis of the eyelid, constriction of the pupil and retraction of the eye (enophthalmos) – Horner's syndrome. For this reason, when performing thoracic sympathectomies for vascular problems in the upper limb, it is very wise to try to avoid the stellate ganglion!

198 Occasionally, the seventh cervical vertebra has a fully developed rib which may articulate with the manubrium sterni. More common, however, is a fibrous cord-like remnant of the cartilaginous rib found in utero. The incidence has been estimated at about 0.5% but it is more important to realise that this condition is one of the factors involved in thoracic inlet syndromes. These may present as vascular insufficiency in the hand or tingling in the fingers due to pressure in the neck on the lower trunk of the brachial plexus.

K Information answers

199 The outer aspect of the pinna is supplied with nerves from two sources. The anterosuperior half is supplied by the auriculotemporal nerve, a branch of the trigeminal (V). The posteroinferior half is supplied by the greater auricular nerve (C2, 3) from the cervical plexus. The back of the pinna is supplied mostly by the lesser occipital nerve (C2). The greater auricular supplies the inferior part adjacent to the lobule. The vagus supplies a small strip of skin posteriorly at the root of the pinna.

For the purpose of description of its nerve supply, the tympanic membrane is divided into anterior and posterior halves. The anterior half is supplied by the auriculotemporal nerve (V). The posterior half is supplied by the vagus (X).

200 The tympanic membrane is connected to the handle of the malleus. The head of the malleus articulates with the body of the incus. The incus in turn articulates, via its long process, with the head of the stapes. The footplate of the stapes rests in the fenestra ovalis. The bones of the tympanic cavity are held in place by ligaments: one runs from the short process of the incus to the posterior wall, another from the anterior process of the malleus to the anterior wall of the cavity. The ossicles act like a gearbox, converting vibrations of the tympanic membrane into in-and-out piston-like movements where the stapes is inserted into the fenestra ovalis (vestibuli).

201 The mastoid antrum lies in front of the sigmoid sinus (venous) and the cerebellum. Infection in the region of the mastoid may spread to involve these structures.

202 The most obvious feature of the medial wall of the middle ear is a convex bulge termed the promontory. This is caused by the underlying first turn of the cochlea. The fenestra ovalis (fenestra vestibuli) lies superoposterior to the promontory. It is for articulation with the footplate of the stapes. Posteroinferior to the promontory lies the fenestra rotunda (fenestra cochleae) covered by its membrane. A ridge may be seen running posteriorly from the region above the promontory to the posterior wall of the tympanic cavity. The ridge is caused by the underlying facial nerve (that part distal to the geniculate ganglion). A ridge of bone is seen running to the area above the promontory from the anterior wall of the middle ear. This ridge of bone supports the tensor tympani muscle. It ends in a pulley (processus cochleariformis) for the tendon to run round in order to pass laterally to the handle of the malleus.

203 The facial nerve enters the petrous temporal bone via the internal acoustic meatus in company with the vestibulocochlear nerve. At the bottom of the meatus it enters a tunnel, the facial canal. The

nerve runs laterally above the vestibule of the inner ear until it reaches the medial wall of the tympanic cavity above the promontory. Here it bends 90° and runs posteriorly towards the posterior wall of the tympanic cavity. A swelling is associated with this abrupt bend – the geniculate ganglion. At the posterior wall it bends again through 90°, this time to turn inferiorly, and running downwards is situated in the medial wall of the aditus. It continues inferiorly along the posterior tympanic wall and leaves the skull at the stylomastoid foramen.

204 The branches of the facial nerve in the petrous temporal bone are the greater petrosal nerve, the nerve to stapedius and the chorda tympani.

The greater petrosal nerve contains preganglionic parasympathetic fibres destined for the pterygopalatine ganglion. From here, postganglionic fibres are distributed to the lacrimal gland and glands of the nose.

The nerve to stapedius is motor to the stapedius muscle, which runs from the pyramid of the posterior tympanic wall to the neck of the stapes. Its contraction 'damps down' movements of the stapes and prevents damage to the inner ear should it be bombarded with loud noises.

The chorda tympani contains two sets of fibres. One set comprises the taste fibres from the anterior two-thirds of the tongue. The other set contains preganglionic parasympathetic fibres destined for the submandibular ganglion where they synapse and are distributed to the sublingual and submandibular salivary glands.

205 The membranous labyrinth is a freely communicating collection of sacs and ducts which lie within the cavity of the bony labyrinth and are separated from them by perilymph. The membranous labyrinth itself contains endolymph.

From anterior to posterior, it consists of the duct of the cochlea, the saccule and utricle and the semicircular ducts. The duct of the cochlea is situated within the bony cochlea. It contains the basilar membrane, which has special hair cells for the appreciation of sound – the organ of Corti. The semicircular ducts lie within the bony semicircular canals. The utricle and saccule are the two central chambers of the membranous labyrinth. They contain the sense organs for the appreciation of the direction of gravity. Their terminations with the utricle contain swellings termed ampullae. It is here that special sense organs are situated to detect acceleration occurring in each of the planes represented by the semicircular canals.

The saccus endolymphaticus is a blind-ended diverticulum of the central portion of the membranous labyrinth.

206 There are three semicircular canals in each inner ear: superior, posterior and lateral. The lateral canal is set in the horizontal plane, its convexity directed laterally. The superior canal is situated vertically, in the axis of the petrous temporal bone which is anteromedial to posterolateral. The posterior canal is also set vertically but at 90° to the petrous temporal axis; therefore, it runs anterolateral to posteromedial. Thus the posterior canal of one ear is set in the same plane as the superior canal of the opposite ear and vice versa.

207 The cochlea consists of a central modiolus around which a hollow cone is wrapped two and a half times. The cochlear nerve ascends in the modiolus and is distributed to each successive turn. The duct of the cochlea is triangular-shaped in cross-section. Its apex is attached to the spinal lamina and its base to the middle third of the outer surface of the bony cochlea. It is filled with endo-lymph. Its inferior margin is the basilar membrane which supports the organ of Corti. Above and below the duct of the cochlea are two triangular spaces – the scala vestibuli and scala tympani, respectively. They are filled with perilymph and are in communication over the apex of the cochlear duct. The scala vestibuli ends in the fenestra ovale on which the stapes impinges. The scala tympani ends in the secondary tympanic membrane over the fenestra rotunda.

K Applied answers

208 The external ear consists of the auricle and external auditory meatus. The auricle is composed of fibrocartilage covered closely by skin. The lobule, or ear lobe, is the only part of the auricle with intervening connective tissue, and is often stretched to accommodate earrings, plugs or other traditional finery. The most obvious features of the auricle are the

outer rounded lip – the helix – surrounding the fossae, which in turn give way to the inner lip or antihelix. Protecting the entrance of the external auditory meatus is a prominent projection, the tragus. The auricular cartilage is continuous with the cartilaginous lateral third of the canal itself. One should note that, at birth, the external auditory meatus is entirely cartilaginous except for the roof. In the adult, however, the medial two-thirds is part of the temporal bone. The entire canal is lined by skin which is closely applied to the bony canal but is separated from the cartilaginous portion by subcutaneous tissue containing numerous ceruminous glands which produce wax.

209 The tympanic membrane is situated at the medial end of the external auditory meatus, separating it from the ossicles of the middle ear. It is a roundish structure facing downwards and forwards, but the effect of parallax makes it appear oval when seen through the otoscope. The normal membrane is glistening and semi-opaque, and the handle of malleus can be seen attached to its deep surface. Occasionally, posteriorly, the incus shadow can be visualised. From the lateral process of the malleus the anterior and posterior malleolar folds form the margin of a lax membrane, the pars flaccida. The remaining membrane is taut and concave where the handle of the malleus pulls it inwards. This area is called the pars tensa and the point of maximum concavity, the umbo. On examination, a reflected 'cone of light' from the otoscope is seen in the anteroinferior part of the membrane.

210 The external auditory meatus is S-shaped; the first part passes forwards and upwards, the second is slightly backwards and the longest third part runs forwards and slightly downwards. To straighten out this canal in the adult, gentle traction is applied in an upwards and backwards direction. In the infant, pulling downwards and backwards is more rewarding.

211 The nerve supply of the canal and tympanic membrane, especially in the posterior and inferior portions, is via the tiny auricular branch of the vagus nerve. This nerve may also carry glossopharyngeal and facial nerve fibres, and some people who are particularly sensitive to stimulations of these nerves may well develop an ear-cough

reflex or, occasionally, vomiting, which may be associated with the vagal and glossopharyngeal innervation. Sometimes even a persistent lump of wax against the tympanic membrane may be the cause of an unexpected cough.

212 Ramsay Hunt syndrome is a herpes zoster infection of the geniculate ganglion, also involving cranial nerves IX and X, which causes vesicles to appear in the distribution of the auricular branch of the vagus nerve. This, therefore, suggests that there are general sensory fibres of the facial nerve carried with the vagus branches to the auricle and external auditory meatus. In this syndrome one sometimes also finds vesicles on the palate and pharynx.

213 Hyperacusis is seen when the stapedius muscle of the middle ear is not functioning properly and cannot therefore damp down the stapes vibrations. An unusual taste in the mouth may have many causes, though damage to the chorda tympani with abnormal salivary secretions is a possible cause. Facial weakness at once makes one think of a VIIth cranial nerve lesion, which could also explain the first two symptoms if the lesion is proximal to the middle ear before the nerve to stapedius and the chorda tympani branch. Vertigo has various causes – both labyrinthine and cerebral – but taken together with the deafness, a likely diagnosis would be a lesion affecting both the VIIth and VIIIth cranial nerves. Clinically, these occur quite often as a tumour in the cerebellopontine angle, arising from the sheath of the vestibular portion of the VIIIth nerve (acoustic neuroma). Other cranial nerves may also be involved, especially the VIth nerve presenting as a lateral rectus palsy.

214 In children an untreated otitis media may spread to the mastoid air cells. From here the infection may well break through the superior wall of the mastoid and into the cranial cavity, affecting the meninges and temporal lobe, and causing meningitis or a temporal lobe abscess. If the posterior wall of the mastoid antrum is invaded, the sigmoid dural venous sinus may be infected, causing a thrombosis with all its very serious consequences. Meningitis, cerebral abscess or venous sinus thrombosis could all cause the symptoms and signs described.

215 Commercially it is good for business but, anatomically, the reason is that swallowing opens up the

auditory tube (Eustachian, pharyngotympanic). The auditory tube allows equalisation of pressure from the middle ear to the outside world. When taking off, and particularly when descending in an aircraft, the pressure changes are quite noticeable, and unless the tube is opened the increased air pressure inside the middle ear may cause quite considerable pain. This also explains why flying with a cold can be uncomfortable, as the mucous secretions may block the auditory tube, causing increased pressure within the middle ear. The exact mechanism of opening the tube is debated, but during swallowing the pharyngeal walls are raised and one of the contributing muscles is salpingopharyngeus, which originates from the pharyngeal end of the cartilaginous tube.

UPPER LIMB

A Information answers

216 The two teres muscles, running superolaterally, and the long head of triceps, parallel to the medial border of the shaft of the humerus, together form the medial triangular space (through which passes the subscapular artery), the lateral triangular space (radial nerve, profunda brachii artery) and the quadrangular space (axillary nerve and posterior circumflex humeral artery).

217 The axillary sheath is a fibrous sheath which encloses the first part of the axillary artery and the adjacent parts of the axillary vein and brachial plexus. It is a prolongation of the prevertebral layer of the deep cervical fascia.

218 The axilla is the shape of a triangular pyramid with its top cut off. The apex of the axilla is situated superiorly and is in continuity with the neck. Its boundaries are: medially the first rib, anteriorly the clavicle and posteriorly the superior border of the scapula. The axillary blood vessels and brachial plexus enter the upper limb through this triangular space. The walls of the axilla are posterior, anterior, lateral and medial. The posterior wall consists of the scapula, covered by the subscapularis above and latissimus dorsi and teres major below. The anterior wall is formed by the pectoralis major and minor,

and clavipectoral fascia where the latter muscle is absent. The medial wall consists of the upper four ribs and intercostal spaces, filled by the intercostal muscles, and the upper half of the serratus anterior muscle. The lateral wall consists of the floor of the bicipital/intertubercular groove with the tendon of the long head of biceps running upwards into the capsule of the shoulder joint. The base is formed by skin and a thick layer of fascia (the axillary fascia) sealing the walls. The axilla contains the axillary vessels, the lower portion of the brachial plexus and its branches, lymph nodes and fat.

219 The medial cord of the brachial plexus is situated medial to the second part of the axillary artery. Its branches are the medial pectoral nerve and the medial cutaneous nerves of the arm and the forearm. It terminates by dividing into two – the ulnar nerve and the medial root of the median nerve. The posterior cord lies posterior to the second part of the axillary artery. Its branches are the upper and lower subscapular nerves and the thoracodorsal nerve (to latissimus dorsi). It ends by dividing into two – the axillary and radial nerves.

220 The brachial plexus is formed from the anterior primary rami of C5–T1. In the neck the trunks are formed, the upper from the union of C5 and C6 roots, the lower from the union of C8 and T1, and the middle by the continuation of C7. Behind the clavicle each trunk divides into two – an anterior and posterior division. All the posterior divisions unite to form the posterior cord. The anterior divisions of the upper and middle trunks join to form the lateral cord, and the anterior division of the lower trunk continues as the medial cord. The cords of the brachial plexus (lateral, posterior and medial) derive their names from their relationship to the second part of the axillary artery, defined as that part overlaid by pectoralis minor, around which they are disposed.

221 The area of skin supplied by a single spinal nerve constitutes a dermatome. Dermatome C5 is situated on the lateral aspect of the shoulder and arm, extending as far inferiorly as the elbow joint. The C6 dermatome is situated on the lateral aspect of the forearm. It includes the thumb and index finger.

222 The female breast is situated in the superficial fascia of the front and side of the chest. Its usual surface marking is from the second to sixth rib vertically and from the sternum medially to the mid-axillary line laterally. It therefore overlies pectoralis major, part of serratus anterior and a small portion of the external oblique. The breast is supported by fibrous septa which run from the fascia overlying the gland into its depths – the suspensory ligaments (Cooper).

223 The lymphatic vessels of the breast originate in the subcutaneous subareolar plexus and a deep plexus overlying the muscle bed. From here the lymphatics run with the blood vessels which supply the gland. Thus some 75% of the lymph runs to the axilla (in the same direction as the lateral thoracic artery) and drains to the pectoral (anterior) nodes, and thence to the apical axillary nodes and subclavian trunk. Most of the remainder of the lymph runs with the perforating branches of the internal thoracic artery to the parasternal nodes adjacent to the latter artery and deep to the rib cage. Pathologically, blockage of lymphatics by tumour tissue can cause anastomoses to open up and allow drainage to other areas, for instance, to the opposite breast or the inguinal region.

224 The serratus anterior, used in punching, protracts the scapula anterolaterally across the lateral thoracic wall, relatively lengthening the reach of the upper limb. It acts as an anchor for this bone, stabilising it and permitting other muscles to act from it on the humerus.

225 The abundant arterial supply of the breast derives mainly from the internal thoracic artery via its perforating branches, which pierce the second, third and fourth intercostal spaces. Also, branches from the axillary artery – the lateral thoracic and thoracoacromial arteries – together with the lateral and anterior cutaneous branches of the third, fourth and fifth intercostal arteries, provide a significant share.

226 The brachial plexus is said to be prefixed when formed by the ventral rami C4–C8, and postfixed when formed from C6 to T2. In the latter type, the inferior trunk may be compressed by the first rib,

producing predictable neurovascular symptoms in the upper limb.

A Applied answers

227 Breaking one's fall from a tree by grasping a branch, or violently pulling an infant's arm during an awkward delivery, may damage the inferior trunk of the brachial plexus, with resultant loss of function at the wrist and digits imparted by the ulnar nerve (C8 and T1 myotomes). Paralysis of the medial part of flexor digitorum profundus reduces power grip while paralysis of the ulnar innervated intrinsic muscles of the hand significantly impairs fine manipulation (Klumpke's paralysis). The 'claw-hand' is the long-term outcome, with intermetacarpal guttering and inability of the long digital flexors to grasp an object, owing to the impaired essential synergistic action of the interossei and medial lumbricals. Loss of sensation may also occur along the ulnar side of the hand, forearm and arm.

228 The scapular anastomosis, connecting the first part of the subclavian (via the thyrocervical trunk) with the third part of the axillary (via the subscapular artery), enables the blood supply to the upper limb to be uncompromised in all positions of the shoulder joint. Accurate identification of the subscapular artery is therefore needed, since the axillary artery must be ligated above it to ensure a continued blood supply to the upper limb.

229 Axillary wounds often involve the axillary vein, owing to its large size and exposed position in the medial aspect of the axilla. Lacerations to its superior part, where it is largest, are particulary dangerous, not only from the risk of profuse haemorrhage, but also from the possibility of air embolism.

230 The skin of the upper limb is drained by the lymphatics in a similar manner to the venous drainage. The skin of the fingers is therefore likely to drain via the dorsal digital lymphatics along the forearm to the cubital fossa, where the majority of vessels follow the basilic vein, through the deep fascia and into the lateral groups of axillary nodes.

231 A cervical rib, in some cases, may cause pressure on the lowest trunk of the brachial plexus or, more commonly, on the T1 nerve root itself. The T1 myotome is sole supply to the intrinsic muscles of the hand, and thus wasting of these muscles may occur in cervical rib cases. The dorsal interossei when atrophied cause a noticeable 'guttering' of the fingers.

232 An Erb's palsy is classically due to an upper brachial plexus trunk lesion, affecting C5 and C6 nerve roots. The damaged arm is held in the 'waiter's tip' position with the shoulder adducted, the elbow extended and the hand pronated. This can be explained by damage to the axillary nerve (C5, 6) causing an adducted shoulder, as well as damage to the musculocutaneous nerve affecting biceps and brachialis causing loss of elbow flexion and power supination of the forearm.

233 A 'winged' scapula is seen in some thin women, but following a radical mastectomy one must immediately think of a weak serratus anterior muscle. The most likely cause is trauma to the long thoracic nerve (Bell) which lies along the medial axillary wall, sending twigs to each of the eight slips of the muscle. It is here that breast surgeons may traumatise it while clearing lymph nodes from the axillary fat. The resulting paralysis makes the patient unable to abduct the arm farther than the horizontal position, since the scapula cannot be rotated to raise the glenoid cavity.

234 If the whole brachial plexus were destroyed, no fibres from C5 to T1 would be capable of sensory appreciation. The bulk of the upper shoulder, however, is supplied by C3 and C4 and these dermatomes are thus unaffected. Remember the distribution of referred pain from the diaphragm.

235 The axillary nerve arises from the posterior cord of the brachial plexus and divides at the lower border of teres minor into an anterior and posterior branch. The anterior branch, accompanied by the posterior circumflex humeral vessels, winds posteriorly through the quadrangular (quadrilateral) space and round the surgical neck of the humerus to supply deltoid from its deep surface. This is usually some 6–8 cm inferior to the bony prominence of the acromion. Any injections below and

posterior to the midpoint of the acromion will endanger the axillary nerve. Muscular and financial incapacity may result!

236 Tethering is due to shortening of the suspensory ligaments (Cooper) which normally fix the breast tissue to skin and underlying fascia. Tumours within the breast may cause shortening and thus pull the skin inwards, creating a dimple.

237 Peau d'orange is due to cutaneous lymphatic oedema and is a classic sign of advanced carcinoma of the breast. It is due to a blockage in the normal lymphatic drainage from the breast, mainly to the axillary nodes, and can even be seen in chronic breast abscesses.

238 The whole breast is enlarged, the axillary tail often being noticeable for the first time, and many dilated veins can be seen over the breast skin. The nipple and areola are well pigmented and normally there is a greatly increased number of areolar glands (sebaceous glands of Montgomery). These glands also increase in size, as their function is to lubricate the nipple during lactation.

239 The most likely path for an areolar condition to spread is via the subareolar plexus to the anterior (pectoral) axillary lymph nodes. Often some large lymphatic vessels can be seen to drain from the nipple along the level of the third rib to the axilla. However, lymphatic spread may sometimes involve the parasternal nodes or the opposite breast and even the mediastinal, peritoneal or inguinal nodes.

B Information answers

240 The axillary artery commences at the lateral border of the first rib as the continuation of the subclavian artery. It ends at the inferior border of the teres major muscle, continuing as the brachial artery. It is divided into three parts by the pectoralis minor muscle. The first part is proximal to the muscle, the second part is behind the muscle and the third part distal to it. Conveniently, the first part has one branch, the second part two and the third part three branches. These are the supreme thoracic from the first part, the thoracoacromial and lateral thoracic from the second part, and the anterior and posterior

circumflex humeral and subscapular artery from the third part. The subscapular artery is the largest branch. The anastomoses around the scapula bring the subclavian, axillary and upper intercostal arteries into contact. They are necessary to allow adequate perfusion of the upper limb in all positions of the shoulder, when occlusion of the main trunk may occur. The subclavian artery gives two branches to the anastomosis – the suprascapular artery and the deep branch of the transverse cervical. These anastomose on the scapular, with the subscapular branch of the axillary artery and with the posterior intercostal arteries over which the scapula lies.

241 The clavicle can be divided into a medial two-thirds (convex anteriorly) and a lateral third (convex posteriorly). The medial end is rounded for articulation with the sternum, the lateral end flattened for articulation with the acromion. The superior surface of the bone is relatively smooth, but the inferior aspect is roughened by various features. Laterally is a ridge for the attachment of the trapezoid ligament, ending in the tubercle for the conoid ligament. Furthermore, there is a distinct groove for the costoclavicular ligament. Using these criteria, the side of a clavicle may be easily ascertained.

242 Two movements are associated with abduction of the arm to 180° – abduction of the glenohumeral joint and rotation of the scapula and entire upper limb. Abduction of the glenohumeral joint is limited to 90°, being initiated by supraspinatus and continued by the middle fibres of deltoid. In order to prevent the head of the humerus being pulled bodily upwards by deltoid during this movement, the teres muscles, subscapularis and infraspinatus contract and hold the head down in the glenoid. Rotation of the scapula is performed by courtesy of a muscular couple consisting of the upper fibres of trapezius, pulling upwards on the scapular spine and serratus anterior pulling on the inferior angle.

243 The coracoclavicular ligament connects the clavicle to the coracoid process of the scapula. It consists of two parts – a flat quadrilateral trapezoid part and a thick triangular conoid part. It helps maintain the articulation of the weak acromioclavicular

joint. Without its help, the clavicle dislocates superiorly and the scapula is displaced medially towards the midline.

244 The muscles associated with the intertubercular (bicipital) groove are the long head of biceps, latissimus dorsi, teres major and pectoralis major. As its name suggests, the groove is occupied by the tendon of the long head of biceps. On the lateral lip of the groove the pectoralis major is inserted. Medially the latissimus dorsi is inserted into the floor of the groove and teres major into the medial lip. There is a mnemonic for this which reinforces the above information: 'Lady Dee (latissimus dorsi) lies in bed between two majors' (pectoralis major and teres major).

245 When considering the factors contributing to the stability of any joint, it is as well to run through a 'checklist' of structures which may be important. These are, first, bony articulations, second, ligaments (intrinsic or capsular and extrinsic or extracapsular) and, third, muscles. Applying this to the shoulder, it will be obvious that, unlike the hip, the shallow ball-and-socket arrangement of the shoulder is very unstable, even allowing for the deepening effect of the glenoid socket by the cartilaginous labrum glenoidale. The capsule of the shoulder is lax, so the intrinsic (glenohumeral) ligaments must also be lax and therefore ineffective in preventing dislocation. The extrinsic coracoacromial ligament, however, forms an arch above the joint, which prevents upward dislocation. The major factor in maintaining stability of the shoulder is the tonus of the muscles which act across it. These are the rotator cuff muscles (infraspinatus, supraspinatus, subscapularis and teres minor) which attach the humeral head to the scapula, and the long head of biceps which acts as a strong strap-like support over the head of the humerus. It should be said that these muscular supports are deficient inferiorly and it is in this direction that the humeral head usually dislocates, with only the tendon of the long head of triceps stabilising the joint inferiorly in abduction. The shoulder joint pays for the large range of movement available to it by being rather unstable.

246 The rotator cuff muscles are the supraspinatus, infraspinatus, teres minor and subscapularis. They

attach the scapula to the head of the humerus. Supraspinatus runs from the supraspinous fossa on the posterosuperior aspect of the scapula to the upper facet of the greater tuberosity of the humerus. Infraspinatus arises from the infraspinous fossa and inserts onto the middle facet of the greater tuberosity. Supraspinatus and infraspinatus are supplied by the suprascapular nerve. Teres minor arises from the upper part of the lateral border of the scapula and inserts onto the lower facet. It is supplied by the axillary nerve. Subscapularis arises from the subscapular fossa on the anterior of the scapula. Whereas the other muscles run above or posterior to the shoulder and a crowded insertion on the greater tuberosity, subscapularis runs anterior to the joint and is the only muscle to insert onto the lesser tuberosity. It is supplied by the subscapular nerves. The rotator cuff muscles are the major factor in determining the stability of the shoulder joint, and run anteriorly, superiorly and posteriorly to it. They are deficient inferiorly, which is the usual direction for shoulder dislocations to occur.

B Applied answers

247 Anterior shoulder dislocation is common in athletes, when excess extension and lateral rotation of the humerus drives the humeral head anteriorly, usually stripping the fibrous capsule and glenoid labrum from the anterior aspect of the glenoid cavity. A hard blow to the humerus with the joint abducted to 90° forces the humeral head inferiorly through this weak, unprotected part of the capsule, where it comes to lie inferior to the glenoid cavity, encroaching on the quadrangular space with potential damage to the axillary nerve. The strong flexor and abductor muscles usually pull the head anterosuperiorly into a subcoracoid position. Unable to use the arm, the patient commonly supports it with the other hand.

248 The clavicle articulates at two atypical synovial joints: both have their articulating surfaces covered with fibrocartilage. The sternoclavicular joint moves like a ball-and-socket arrangement, despite the saddle-like form of its articulating surfaces: it moves through 60° on full arm abduction! It depends entirely on its ligaments and articulating

disc for its stability. The fibrous disc completely partitions the joint into two synovial cavities, a typical adaptation where two planes of movement occur simultaneously at a joint (cf. the temporomandibular and knee joints). However, the disc, by its attachments, also has a role in preventing medial displacement of the clavicle superior to the manubrium.

The strong extrinsic costoclavicular ligament limits ascent of the medial end of the clavicle (e.g. when carrying a heavy suitcase). The acromioclavicular joint is a plane synovial joint, with the clavicle tending to override the acromion. Its articular disc is incomplete, projecting from the superior aspect of the capsule. A hard fall onto the shoulder can dislocate this joint; seriously when the strong coracoclavicular ligament is ruptured as well as the acromioclavicular ligament.

249   The effects of a partial or complete tear in the supraspinatus tendon is pain and spasm in the middle range of abduction – 60°–120°. This is called the 'painful arc syndrome'. In a complete tear the patient cannot abduct the arm without shrugging the shoulder. Deltoid alone will not fully abduct the arm.

250   If the surgical neck is damaged or the shoulder dislocated, then the axillary nerve is endangered as it winds posteriorly from the brachial plexus to the deep side of the deltoid muscle. A damaged axillary nerve may result in a wasting of deltoid with a square-shaped shoulder as well as a small sensory loss on the upper lateral part of the arm.

251   The anatomical neck is the old epiphyseal plate and surrounds the smooth articular cartilage of the head of the humerus. The surgical neck, however, is some 2–3 cm below the head and greater tuberosity and is the site of fracture during injury. It is the most superior part of the shaft.

252   A fall on an outstretched arm often causes this injury, which is one of the commonest sites of fracture in the body. The lateral fragment is depressed due to gravity and the weight of the limb but it is drawn medially and anteriorly by teres major, latissimus dorsi and, especially, pectoralis major. The eversion of the medial fragment is due to sternocleidomastoid.

253 An axial X-ray in abduction is especially useful to demonstrate a dislocated shoulder. This view will distinguish between the common anterior dislocation and the rare posterior dislocation.

254 The term 'shoulder separation' is a misnomer. It refers to the dislocation of the acromioclavicular joint from a hard fall onto the shoulder.

C Information answers

255 This question is a little mean, as every student knows the axiom that 'veins are variable'. Nevertheless, in practice, the veins of the upper limb are used to take blood, to set up infusions and to pass long cannulae to the central veins and heart. The deep fascia separates the veins into a superficial network superficial to it and a deep network deep to it. The superficial veins start in the dorsal plexus on the back of the hand. Two relatively constant veins ascend the arm from this point – the cephalic vein laterally and the basilic vein medially. The cephalic vein crosses the anatomical snuffbox, runs laterally up the forearm and arm, and then ascends in the furrow between the deltoid and pectoralis major muscles (deltopectoral groove). It pierces the clavipectoral fascia to drain into the axillary vein. The basilic vein runs up the medial side of the forearm. Halfway up the arm it pierces the deep fascia. At the lower border of teres major it becomes the axillary vein. There is usually a communication between the cephalic and basilic veins in the cubital fossa – the median cubital vein. The deep veins of the upper limb accompany the arteries in pairs as the venae comitantes. They drain to the venae comitantes of the brachial artery and from there to the axillary vein.

256 The median nerve is formed by two roots: a lateral root from the lateral cord of the brachial plexus and a medial root from the medial cord. Initially the nerve lies lateral to the brachial artery. However, halfway down the arm it crosses anterior to the artery to lie medial to it. In this position they descend together into the cubital fossa where they run deep to the bicipital aponeurosis (grâce à Dieu fascia). The median nerve leaves the cubital fossa by passing between the two heads of pronator teres.

The brachial artery divides near the apex of the cubital fossa into the radial and ulnar arteries. The ulnar artery passes deep to pronator teres to leave the cubital fossa; the radial artery leaves it by passing below brachioradialis.

257 Three nerves are related directly to the humerus at some point, and may be damaged by fractures at these points. The axillary nerve runs round the posterior part of the surgical neck of the humerus. Here it may be damaged by fractures or shoulder dislocations. The radial nerve runs round the spiral groove of the humeral shaft posteriorly. The ulnar nerve runs posterior to the elbow joint in a groove between the medial humeral epicondyle and olecranon of the ulna, where you 'hit your funny bone'.

258 The cephalic vein runs up the lateral aspect of the arm to enter the depression between the pectoralis major and deltoid muscles (deltopectoral groove). Here it pierces the clavipectoral fascia to enter the axillary vein.

259 When the elbow is fully extended and the hand supinated, the long axis of the ulna is not in a direct line with the long axis of the humerus but makes an angle with it. The forearm is angled laterally on the upper arm at an angle of about 160° – termed the carrying angle. It is caused by the fact that the medial lip of the trochlea extends further inferiorly than the lateral lip. The carrying angle is particularly marked in women.

260 The elbow can be divided into three subjoints, although they all share the same synovial cavity. First is the humeroulnar joint. This is the articulation between the trochlea of the humerus and the upper end of the ulna. It is here that flexion and extension occur. Second is the humeroradial joint. This is between the round capitulum of the humerus and the disc-shaped head of the radius. Third is the superior radioulnar joint. This is the articulation between the head of the radius and the radial notch on the ulna plus the annular ligament which runs round the radial head. The superior radioulnar joint and humeroradial joint allow the head of the radius to swivel against the capitulum. When transferred down the forearm this is converted into pronation and supination.

261 Pronation and supination occur about an axis which is a line drawn from the centre of the radial head to the base of the styloid process of the ulna. Pronation is brought about by pronator teres, running from the common flexor origin and ulna to a tubercle halfway down the radius, by pronator quadratus, joining the ulna and radius in the distal quarter, and weakly by brachioradialis. Supination occurs chiefly by courtesy of biceps which inserts on the posterior aspect of the radial tuberosity, thereby 'unwinding' the radius as it contracts, and to a lesser extent by supinator running from the ulna and humerus posterior to the interosseous membrane and wrapping itself round the proximal part of the radius. They are aided by the two dorsal thumb muscles arising from the ulna, viz. extensor pollicis longus and abductor pollicis longus.

262 The ulnar nerve can normally be palpated as it passes in its groove behind the medial epicondyle. The superficial radial nerve can be palpated as it crosses the tendon of extensor pollicis longus on the dorsum of the wrist.

263 The cubital fossa is the area anterior to the elbow joint. It is triangular in shape with its base proximally and its apex directed distally. Its base is an imaginary line drawn between the two humeral epicondyles. Its lateral margin is formed by the brachioradialis muscle and the medial margin by pronator teres. They cross at the apex. The floor is composed of brachialis with a portion of supinator laterally. The roof is skin, superficial and deep fascia, the latter being reinforced medially by the bicipital aponeurosis. The fossa contains, from lateral to medial, the radial nerve, the biceps tendon, the brachial artery, bifurcating into its two terminal branches, and the median nerve.

264 The anterior interosseous nerve is derived from the median nerve just proximal to pronator teres. It descends on the anterior aspect of the interosseous membrane with the anterior interosseous artery, between flexor pollicis longus laterally and flexor digitorum profundus medially. It ends on the carpus. It supplies the flexor pollicis longus, flexor digitorum profundus and pronator quadratus. It sends articular branches to the wrist, inferior radioulnar and carpal joints.

The posterior interosseous nerve arises from the radial nerve as it crosses the lateral humeral epicondyle. It runs through supinator, curving round the neck of the radius as it does so. It passes between the superficial and deep extensor muscles of the forearm, and at the lower border of extensor pollicis brevis, comes to lie on the posterior aspect of the interosseous membrane. It ends on the carpus. It supplies the extensor muscles of the forearm, wrist and carpal joints.

265 The radial nerve is a branch of the posterior cord of the brachial plexus. It enters the posterior compartment of the arm with the profunda brachii artery under cover of the uppermost fibres of the medial head of triceps, below teres major. It then enters the lower part of the spiral groove between the lateral and long heads of triceps. Here it is in direct continuity with the humerus and may be damaged by pressure or a fracture. In the groove it passes from the medial to the lateral side of the muscular septum, to enter the anterior compartment. It descends between brachialis medially and brachioradialis laterally to reach to cubital fossa.

C Applied answers

266 The course of the brachial artery through the arm is represented by a line connecting the midpoints of the clavicle and the cubital fossa. It is superficial and palpable throughout its entire course. Running on the medial side of biceps brachii, anterior to triceps and brachialis, it reaches the cubital fossa, opposite the neck of the radius. Its compression is, therefore, possible in order to control haemorrhage, the best place being near the middle of the arm, where it lies on the tendon of coracobrachialis, medial to the humerus. To compress it in the inferior part of the arm, pressure is needed posteriorly, as it lies anterior to the bone here. In 20% of cases the brachial artery is doubled in all or in part of its course.

267 Within a few hours, irreversible paralysis of the deep forearm flexors occurs (flexor pollicis longus, flexor digitorum profundus). Eventually necrotic muscle is replaced by fibrous scar tissue to cause their permanent shortening and a flexion deformity known as 'Volkmann's ischaemic contracture'. It

may also be caused by the extended and improper use of a tourniquet.

268 The brachial artery is the direct continuation of the axillary artery distal to the teres major muscle and lies medial to the upper humerus but winds anteriorly in front of brachialis to cross the elbow joint anteriorly between the humeral condyles. Its pulsations are palpable throughout its length as it descends the arm because it is only covered by skin and fascia. In the cubital fossa it is covered and protected by the bicipital aponeurosis, which separates it from the numerous superficial veins of this region. Normally, it divides at the level of the elbow joint. However, it is the commonest large artery which is subject to anomalies such as a high bifurcation, collateral vessels or a double artery, which may envelop the median nerve.

269 'Crutch palsy' is due to pressure on the radial nerve as it leaves the axilla. This results in an inability to form as tight a grip as usual, as for this action, both extensors (radial nerve) and flexors are required to form a tight power grip. This is an example of synergism and can easily be demonstrated by trying to grip something very tightly with your wrist in full flexion. The wrist weakness will be due to loss of power in the finger and wrist extensors.

270 He has torn his biceps tendon, thus causing a large bulge of ruptured muscle fibres on his anterior arm. The pain is most likely to be referred to the shoulder joint and his difficulty in opening a door is due to the loss of biceps' two functions – elbow flexion and, in this case more specifically, supination.

271 This fracture results from a fall on the outstretched arm, and the seriousness of the injury depends on the degree of displacement anteriorly of the proximal humeral fragment. When the proximal fragment displaces, it may cause irritation or laceration of the brachial artery and median nerve. The loss of median nerve function may result in severely impaired pronation, reduced wrist flexion with ulnar deviation and, most drastically, the loss of thumb opposition as well as accompanying sensory deficits. In the long term, ischaemia due to interruption of the brachial artery may result in a

Volkmann's contracture of the hand due to fibrosis of the long flexor and extensor muscles of the forearm. In this condition, the wrist is flexed and the metacarpophalangeal joints are extended whilst the interphalangeal joints flexed.

272 Childhood instability of the elbow joint is due to the late fusion of the epiphyses of the ends of the bones involved (humerus, radius and ulna). The proximal part of the olecranon fuses with the body of the ulna between 16 and 19 years; the head of the radius fuses with its body between 15 and 17 years. Hence separation of these epiphyses can occur from a fall onto the elbow (the epiphyseal cartilaginous plate being weaker than the surrounding bone). Falls in children causing abduction of the extended elbow joint can cause traction on the ulnar collateral ligament with avulsion of the medial epicondyle of the humerus (the epiphysis of the medial epicondyle usually fuses by 14 in the female, 16 in the male). Traction injury of the ulnar nerve frequently complicates this, since it passes posterior to the medial epicondyle between it and the ulnar olecranon.

273 'Students' elbow' is a friction bursitis of the subcutaneous olecranon bursa. Rarely, assembly line workers get subtendinous olecranon bursitis from repeated flexion and extension of the forearm. The radioulnar bursa (between extensor digitorum, the radiohumeral joint and supinator) may become inflamed in backhand tennis players who repeatedly flex and violently extend their wrist with a pronated forearm; they suffer pain on elbow extension with a pronated forearm. Bicipitoradial bursitis causes pain on forearm pronation, as this bursa is compressed against the anterior half of the radial tuberosity by the biceps tendon.

274 Repeated forceful use of the superficial muscles in the anterior aspect of the forearm, as occurs during golfing, strains the common flexor origin at the medial humeral epicondyle with resulting inflammation and pain at that location. Repeated backhand strokes in tennis cause strenuous contraction of the forearm extensors with lateral epicondylitis; typically, pain is increased by grasping the racquet, putting tension on the common extensor origin.

275 Division of the ulnar nerve at the elbow will cause only minimal sensory and vasomotor loss on the medial side of the hand and fingers. The resultant 'claw-hand' is due to paralysis of flexor digitorum profundus of the fourth and fifth fingers and all the intrinsic hand muscles. As these muscles atrophy, a 'guttered' look often appears between the metacarpals. Although there is a slight flexion deformity of the fourth and fifth fingers and the patient cannot abduct or adduct his fingers with his palm flat on the table, it is amazing how this injury may leave a reasonably functioning hand.

276 The child has a dislocated head of radius. This injury is seen most commonly in young children whose annular ligament of the radial head has not yet reached its normal configuration. In the adult it is cone-shaped, encircling the bulbous radial head, whereas in children it is a simple signet-ring shape and the radial head is not developed, making it easy for a sudden pull to sublux the head out of the annular ligament. It is reduced by firm supination, screwing the radial head back into its anatomical position.

277 For a mainly right-handed population a right-handed thread means that supination is the action used to fix a screw. This is a strong movement using mainly the biceps muscle, whereas the pronator teres and pronator quadratus, which are used in unscrewing, are not so powerful. A left-handed thread would reverse this situation.

278 'Miner's elbow' is in fact an olecranon bursitis due to the rubbing of the subcutaneous tissue overlying the olecranon whilst crawling down the pits or writing great volumes of work. It is a disease of the enthusiastic author!

279 There is normally a simple relationship between the epicondyles and the olecranon; in the extended elbow they lie on a straight line, whilst in the flexed elbow they form an equilateral triangle. If the elbow is dislocated these relationships will alter, whereas in supracondylar fracture they remain constant.

280 The surgeon performs a transposition of the ulnar nerve, taking it from just posterior to the medial epicondyle and replacing it on the anterior aspect of the forearm. This usually solves the tingling

problem but leaves the nerve relatively exposed for forearm injuries.

281 The bicipital aponeurosis or 'grâce à Dieu' fascia protects the median nerve and brachial artery from such incidents. The expanding mass is therefore likely to be a haematoma or traumatic aneurysm of the brachial artery due to the student not realising he had probed too deep.

282 Both palmaris longus and peroneus tertius are atavistic muscles not found in all of the population. Palmaris longus is missing in about 5–10% of the population and, therefore, in these people it is often easy to palpate the median nerve prior to its passing through the carpal tunnel.

D Information answers

283 The common extensor origin gives rise to four of the five superficial extensor muscles of the forearm: extensor digiti minimi, extensor carpi ulnaris, extensor carpi radialis brevis and extensor digitorum. It is situated on the lateral epicondyle of the humerus. Note that extensor carpi radialis longus arises from the lateral supracondylar ridge.

284 The wrist joint is a synovial joint of the condyloid variety. The articular surfaces are formed by the scaphoid, lunate and triquetral bones distally. These form a convex cavity proximally which articulates with a concave socket, formed by the lower end of the radius laterally and the triangular fibrocartilage medially. The ulna gives no contribution to the articular surface. The capsule is attached to the bones mentioned above and is reinforced laterally and medially to form the respective collateral ligaments. The movements permitted are flexion, extension, abduction and adduction (in combination, circumduction). Rotation cannot occur because the ellipsoid nature of the opposing surfaces prevents it. The function of rotation is taken over by pronation and supination.

285 Flexor carpi ulnaris (FCU) and flexor carpi radialis (FCR) pass anterior to the wrist joint on the ulnar and the radial aspect, respectively. Similarly, extensor carpi ulnaris (ECU) and extensor carpi

radialis longus (ECRL) and brevis (ECRB) pass posterior to the wrist on their respective aspects of the joint. When the two flexors contract together, flexion of the wrist results, and when the extensors contract, extension. If the ulnar flexor and ulnar extensor now contract, adduction of the wrist results. If the radial flexor and extensor are employed, abduction occurs. Therefore, the abductors of the wrist are FCR, ECRL and ECRB, and the adductors are FCU and ECU.

286 Muscles can only contract a certain proportion of their length. It follows that, if their origin and insertion are approximated by this amount, no further contraction can occur and their action will be weak. If the wrist is flexed, the long flexors of the fingers work under this disadvantage. If, however, the wrist is extended prior to gripping, the origin and insertion of the flexors are moved apart, the flexor muscles are stretched and the resulting grip will be powerful.

287 The anatomical snuffbox is situated on the lateral aspect of the wrist. It is bounded by the tendon of extensor pollicis longus medially and the tendons of abductor pollicis longus and extensor pollicis brevis laterally. Its floor consists of the styloid process of the radius, the scaphoid bone, the trapezium and the base of the first metacarpal from proximal to distal. It contains the radial artery as it runs from the ventral to dorsal aspect of the wrist. The snuffbox is crossed by the cephalic vein and the cutaneous branches of the radial nerve.

D Applied answers

288 A fall onto the outstretched hand may fracture the distal radius. In the common Colles' fracture, the distal radial fragment is displaced posteriorly and impacted, shortening the bone. Normally the radial styloid is palpated approximately 1 cm distal to the ulnar styloid, but their palpation at the same horizontal level indicates bony displacement.

289 Occasionally there are several slender communicating fibres between the median and ulnar nerves in the forearm. Therefore, even in complete median nerve lesions some of the muscles usually affected

will be spared. Also, the ulnar nerve can not uncommonly encroach on the innervation of the thenar muscles, adding to diagnostic confusion.

290 The ulnar artery descends superficial to the flexor muscles in 3% of people. This must be kept in mind when performing intravenous injections in the cubital area, since an accidental intra-arterial injection with certain drugs, when mistaken for a vein, may cause gangrene from partial or total loss of the hand.

291 Because herpes zoster virus infection travels along a single nerve, it often demarcates the dermatomes. The C6 dermatome usually involves the lateral side of the forearm below the elbow and includes the thumb, thenar eminence and index finger. It does not cross the axial line and is always along the preaxial border of the arm, though, as with all dermatomes, its edge is best defined by a smudge and not a definite clear line. Axillary nerve injury (C5) would produce numbness over the 'regimental badge area' on the lateral aspect of the arm over the deltoid tuberosity of the humerus.

292 There is little difference in function between ulnar nerve wrist and elbow injuries, but the wrist injury leaves the patient with a more pronounced claw on the medial fingers. This is due to the fact that the flexor digitorum profundus is not affected and thus causes a more pronounced claw. In an elbow injury to the ulnar nerve, this muscle is paralysed and the claw-hand is therefore less obvious. The sensory loss in both instances may be quite minimal and, surprisingly, it may be difficult to be sure that the nerve is damaged at all.

293 Other than the obvious cut muscle tendons, the two major nerves of the hand are both vulnerable. The median nerve lies just deep to palmaris longus and the ulnar nerve lies between the ulnar artery and pisiform bone. To test the ulnar nerve, the hand is placed palm downwards and the fingers straightened, then abducted and adducted against resistance (or try the paper holding test between adducted fingers). To test the median nerve, the patient is asked to abduct his thumb against resistance (note that thumb opposition can be mimicked by the long flexor tendons). True thumb

opposition is impossible in long-standing median nerve lesions, as the thumb is laterally rotated and adducted and consequently looks like a monkey's or ape's hand.

294 Post-traumatic pain deep in the anatomical snuffbox is usually due to a fractured scaphoid bone. This often does not show up immediately, but after 2 weeks healing will be seen on an X-ray. Due to the blood supply arrangement this bone is a classic site of avascular necrosis.

295 Try this yourself and you will find that the power grip is very weak in acute wrist flexion. Normally, the wrist extensors work synergistically with the flexors of the fingers. Flexion of the wrist deprives the long flexor tendons of the ability to contract further and make a strong grip. Flexing the wrist will therefore make someone relax his grip and drop whatever is in his hand.

E Information answers

296 The bones of the middle finger to be discussed will be the three phalanges (distal, middle and proximal) and the metacarpal. The muscles concerned are the long flexors, the extensor, the second lumbrical, the interossei and the adductor pollicis. The flexor digitorum profundus is inserted into the base of the distal phalanx. The flexor digitorum superficialis splits into two to allow the profundus to pass through it and is inserted into the base of the middle phalanx. The single extensor tendon is enlarged posterior to the metacarpophalangeal joint to form the extensor hood, which sends slips into the bases of all three phalanges. The lumbrical arises from the profundus tendon in front of the metacarpal and, passing lateral to the digit, is inserted into the extensor expansion. The middle finger's metacarpal has no palmar interosseous muscle attached to its anterior border but, instead, receives the transverse head of the adductor pollicis. Two dorsal interossei, however, insert onto the extensor expansion and base of the proximal phalanx of this digit, one on each side. They are bipennate muscles and arise from the adjacent borders of the second and third and third and fourth metacarpals. They move the middle finger either way in the coronal plane.

297 Where tendons run over bones they require lubrication by synovial sheaths. In the case of the long digital flexors of the hand, these conditions are met in the carpal tunnel and in the digits themselves. In the palm, the eight tendons of the flexor digitorum profundus and superficialis share a single sheath. The tendon of flexor pollicis longus has one of its own. Each digit of the hand possesses a digital synovial sheath for the profundus and superficialis tendons to that finger, the one to the little finger communicating with the palmar sheath. The flexor pollicis longus of the thumb has one of its own which connects with the palmar sheath of that tendon.

298 The intrinsic muscles of the hand have their fleshy bellies actually in the hand, i.e. distal to the wrist joint. They are the palmaris brevis, the thenar and hypothenar muscles, adductor pollicis, the lumbricals and the palmar and dorsal interossei. They are supplied by the T1 root via the ulnar and median nerves.

299 The myotome of T1 is distributed to the small muscles of the hand. The T1 dermatome is along the medial side of the elbow and arm.

300 The thenar muscles (flexor pollicis brevis, abductor pollicis brevis and opponens pollicis, but not adductor pollicis) are usually supplied by a branch of the median nerve which takes a recurrent and superficial course from the distal aspect of the flexor retinaculum to supply them. The hypothenar muscles are supplied by the deep branch of the ulnar nerve as it passes between the pisiform and the hook of the hamate. This nerve also usually innervates the adductor pollicis muscle.

301 The interossei are of two sorts – palmar and dorsal. The palmar interossei arise from the metacarpal and are inserted into the dorsal expansion of the same digit. Those from the little and ring fingers arise from the radial aspect, that from the index finger from the ulnar aspect. The middle finger is deficient. The dorsal interossei are bipennate, arising from adjacent metacarpals to be inserted onto the extensor hood and the base of the proximal phalanges – the first two on the radial aspect of the index and middle fingers, the last two on the ulnar aspect of the middle and little fingers. The palmar

interossei adduct the digits towards an imaginary axis along the middle finger; the dorsal interossei abduct them. The dorsal interossei to the middle finger (axis) can move it either way. Together, they flex the metacarpophalangeal joints and extend the interphalangeal joints by virtue of their insertion into the extensor hood.

The lumbricals arise from the flexor digitorum profundus tendons, adjacent to the bases of the metacarpals, and passing laterally along their respective digits, are inserted into the extensor hood. They flex the metacarpophalangeal joints but, by pulling on the extensor hood, extend the interphalangeal joints.

Failure of these muscles results in inability to extend the interphalangeal joints while keeping the metacarpophalangeal joints flexed, a movement essential in writing.

302 The ulnar and radial arteries anastomose in the palm to form a superficial and a deep palmar arch.

The superficial arch is formed mainly by the ulnar artery. It curves across the palm deep to the palmar aponeurosis, level with the distal border end of the fully extended thumb. It sends digital branches (two to each digit, one on each side) to the medial three and one-half digits (little, ring, middle and half the index fingers). The radial artery passes across the snuffbox to the dorsum of the hand, and re-enters the palm by diving through the first interosseous space, to give the main input to the deep palmar arch. It runs across the palm anterior to the metacarpals and interossei at a more proximal level than the superficial arch, being level with proximal border of the extended thumb. The deep branch of the ulnar nerve lies within its curve. It supplies the lateral one and a half digits (thumb and half of the index finger) by digital branches. These branches are named the arteria radialis indicis and the arteria princeps pollicis.

303 The superficial lymph vessels of the upper limb follow the cephalic vein laterally and basilic vein medially. The former end in the infraclavicular nodes of the deltopectoral groove. The latter pass through the epitrochlear node, as the basilic vein pierces the deep fascia halfway up the arm, and end in the lateral axillary nodes around the axillary vein. The deep lymph of the arm also passes to

these nodes. The lateral nodes in turn drain to the apical axillary lymph nodes.

304 The thumb is set at right angles to the other digits. Its movements are therefore expressed differently from those of the other digits and this leads to confusion. When discussing the movements of the digits, a line should be drawn in the plane of the nail of the digit concerned. Movement towards the pulp of the digit is flexion, and towards the nail is extension. Movements in the plane of the nail are abduction and adduction. In the case of the thumb, abduction brings the whole digit anteriorly in the anatomical position; movement posteriorly is adduction. Flexion of the thumb at the carpometacarpal joint is associated with medial rotation. In this way the pulp of the thumb is brought into apposition with the pulp of another digit. This is termed opposition. Extension is similarly associated with lateral rotation.

E Applied answers

305 Sudden tension on a long extensor tendon may avulse part of its attachment to the distal phalanx. This can occur on sudden hyperflexion of the distal interphalangeal joint ('cricket finger'). The patient is unable to extend the distal interphalangeal joint.

306 Wrist drop is the most noticeable effect, due to loss of all finger and wrist extensors. This also considerably weakens the power grip. The sensory loss, however, is often only very limited due to overlap of the nerves' sensory distribution but may include a strip down the posterior aspect of arm and forearm as well as a small area of anaesthesia over the first dorsal interosseous muscle.

307 The median nerve lesion is the most disabling because thumb opposition is lost, as well as the sensation over the thumb, index and middle fingers. Consequently, all fine pincer movements such as writing are almost impossible.

308 Dupuytren's contracture is due to fibrosis within the palmar aponeurosis causing flexion of the metacarpophalangeal and proximal interphalangeal joints, especially in the ring and little fingers. The distal interphalangeal joint is not

affected, as the palmar aponeurosis inserts only into the middle and proximal phalanges.

309 The small muscles of the thumb are all innervated by the median nerve, which has to pass through the carpal tunnel below the flexor retinaculum. The skin above the muscles, however, is supplied by the superficial palmar branch of the median nerve, and this branches off just proximal to the retinaculum and thus escapes compression.

310 The long flexor tendons of the thumb are surrounded by synovial sheaths. An infection in the pulp might readily spread proximally along the flexor sheath where, in some people, it joins the ulnar bursa, both of them passing deep to the retinaculum and extending for 2–3 cm to the level of the wrist.

311 Guttering of the dorsum of the hand is seen in atrophy of the interossei. Froment's sign is seen when a patient is asked to grip a piece of paper between thumb and index finger, and due to adductor pollicis paralysis, the patient compensates by visibly contracting flexor pollicis longus instead of adducting the thumb. Clawed ring and little fingers, as with the other signs, would suggest an ulnar nerve lesion.

312 This operation is referred to as a cervical sympathectomy, when in fact it is an upper thoracic sympathectomy performed through an incision in the neck. Cutting the sympathetic chain below the T3 ganglion, one tries not to touch T1 as this may result in a Horner's syndrome. The removal of T2–4 sympathetic segments removes control of sweating and vasoconstriction and, hopefully, results in a warmed dry hand, instead of the painful cold sweating hand seen in Raynaud's disease.

THORAX

A Information answers

313 The commoner variations in sternal anatomy include a sternal foramen, a result of abnormal ossification (not to be confused with a bullet-hole on chest X-ray!), pectus excavatum ('funnel chest'), where the sternal body projects inferoposteriorly

and presses on the heart (which appears enlarged on an AP chest X- ray, with a lateral view revealing all!), pectus carinatum ('pigeon chest'), where the chest is flattened on each side and the sternum projects anteriorly, and the rarer 'cleft sternum', where the fetal halves of the sternum fail to unite owing to defective ossification.

314 In the newborn, collagen and elastic fibres join the two bones, followed by union with hyaline cartilage, which later forms a secondary cartilaginous joint with fibrous tissue uniting the two cartilaginous plates. In 10% of people over 30 years, the superficial part becomes ossified, and in the very old, bony fusion of the sternal body and manubrium may occur (synostosis).

315 All the intercostal muscles (external, internal and innermost) aid inspiration by elevating the ribs, expanding the thoracic cavity in both the transverse and anteroposterior planes. They keep the intercostal spaces rigid, preventing sucking in of the pleura and lung tissue during inspiration and their blowing out during expiration. The external intercostal muscles run inferoanteriorly from as far posteriorly as the rib tubercles, being replaced anteriorly by the external intercostal membrane at the costochondral junctions. The deeper internal intercostal muscles run perpendicular to the externals, inferoposteriorly from the floor of the costal groove of the rib above, to the blunt superior border of the rib shaft below, from the sternum to the rib angles, being replaced there by the internal intercostal membrane. The innermost intercostal muscles are really the deeper part of the internals at the middle of the rib shafts, providing a plane for the intercostal neurovascular bundles to course through.

316 The branches of a typical intercostal nerve are (1) sympathetic rami communicantes (the nerve sends white, and receives grey) (2) a collateral branch, arising near the rib angle, that runs at a lower level than the main nerve close to the superior border of the rib below, supplying the intercostal muscles (3) a lateral cutaneous branch arising beyond the angles, piercing the muscles halfway round the thorax and dividing into anterior and posterior branches that supply skin over lateral aspects of thoracic and abdominal walls (4) an anterior cuta-

neous branch, and (5) muscular branches to the subcostals, transversus thoracis, levatores costarum and serratus posterior.

317 In the first part of their course, the first and second intercostal nerves pass on the internal surfaces of their respective ribs. The first has no anterior or lateral cutaneous branches, and divides into a large superior branch that joins the brachial plexus, and a smaller inferior branch that continues as the first intercostal nerve. The second may also contribute a small part to the brachial plexus, and its lateral cutaneous branch is named the intercostobrachial nerve as it supplies the axillary floor.

318 The first two posterior intercostal arteries arise from the superior intercostal artery, a branch from the costocervical trunk of the subclavian artery, while the 3rd–11th and the subcostal arteries arise posteriorly from the thoracic aorta. Running anteriorly through the intercostal space (at first between the internal intercostal membrane and the pleura and then between the innermost and internal intercostal muscles) they enter the costal groove close to the rib angle, lying between the vein above and the nerve below. Two anterior intercostal arteries are given for each space, the upper six from the internal thoracic artery (from the first part of the subclavian artery) while the 7th–9th originate from one of its terminal branches, the musculophrenic artery. Note that the lowest two spaces have none, being totally supplied by the posterior intercostal arteries.

319 Pleural reflections are the abrupt lines where parietal pleura changes direction from one wall of the pleural cavity to another. The sternal lines of pleural reflection pass inferomedially from the sternoclavicular joints to meet (or even overlap) at the anterior median line at the angle of Louis (T4). On the right, it descends in the median plane and turns left, posterior to the xiphoid process. On the left, on reaching the fourth costal cartilage, it passes to the left margin of the sternum and descends to the sixth costal cartilage. Bilaterally, the costal line of pleural reflection passes obliquely across the eighth rib at the mid-clavicular line, the tenth rib at the mid-axillary line and the twelfth rib at its neck (or just inferior to it).

320 In the upper part of the thorax, during inspiration, the first rib remains relatively fixed. Owing to the anteroinferior slope of the ribs, contraction of the external intercostal muscles forces the sternum forwards and upwards like the handle of a pump, thus increasing the anteroposterior diameter of the thorax.

In the area of the lower ribs, contraction of the external intercostals raises the centre of the ribs as one would lift the handle of a bucket, thereby increasing the transverse thoracic diameter.

321 The following 'events' occur at the plane of T4:
Manubriosternal joint (angle of Louis)
Beginning and end of aortic arch
Bifurcation of trachea
Junction of superior and inferior mediastinum
Second costosternal joint
Confluence of vena azygos with superior vena cava
Thoracic duct runs from right to left
Ligamentum arteriosum lies on this plane.

322 The lower five pairs of ribs are 'false ribs'; that is, they do not articulate directly with the sternum. The eleventh and twelfth ribs have no anterior articulation and are the 'floating ribs'. The eighth, ninth and tenth pairs are attached to each other via their costal cartilages, forming the costal margins.

323 There are two articulations with the vertebral column.

In the costovertebral joint, the head of the rib articulates via synovial joints with the bodies of T6 and T5 and via a ligament, with the intervening disc.

In the costotransverse joint, the tubercle of the rib articulates with the right transverse process of T6 via a synovial joint; it is strengthened by associated ligaments which are the two from the transverse processes of T6 (costotransverse ligament and lateral costotransverse ligament) and one to the transverse processes of T5 (superior costotransverse ligament).

A Applied answers

324 The pleurae descend inferior to the costal margin in three regions, where a surgical incision might

inadvertently enter a pleural sac: the right part of the infrasternal angle, and the right and left costovertebral angles.

325 The angles of the ribs are their most posterior and weakest parts. Crushing injuries tend to break ribs just anterior to their angles, whereas a direct blow may fracture a rib anywhere, the broken ends possibly being driven inwards and lacerating internal organs (e.g. the heart, the lungs or the spleen). The ribs of children, being mainly cartilaginous, are rarely fractured. Multiple fractures in the anterolateral thoracic wall allow a loose segment of wall to move paradoxically with respiration, impairing ventilation ('flail chest').

326 These three points are all bony landmarks so they are relatively consistent at T2, T4/5 and T7 vertebral levels, respectively. The manubriosternal joint is also called the sternal angle (Louis). It is an important landmark for counting ribs, as the second costal cartilage articulates laterally at this joint. In most normal adults it is fairly easy to palpate and is thus commonly used by clinicians as a reference point.

327 Lymph from all of the skin and superficial fascia of the back and abdomen above the level of the umbilicus and iliac crests drains to the axillary group of lymph nodes. Structures in the skin of the chest drain to the anterior axillary nodes, and thus a melanoma of the back would drain to the posterior axillary lymph nodes. These are found around the subscapular vessels lying on the subscapularis muscle and are consequently often termed the subscapular group of axillary lymph nodes. Metastases might spread from this group to the central groups of nodes, in the axillary fat, and eventually to the apical nodes which lie on the axillary vein between pectoralis minor and the clavicle.

328 The level of the umbilicus is a watershed for lymphatic drainage of the skin of the anterior abdominal wall, so a truly paraumbilical tattoo will drain both above and below this watershed. In practice, this means that both right and left anterior axillary nodes, as well as both groups of superficial inguinal nodes, are likely to be affected.

329 However effective a single spinal nerve block is, it is most unlikely to cause any sensory loss because the overlap between contiguous dermatomes is quite considerable. In fact, each segmental nerve overlaps its neighbour to such a degree that, unless at least two consecutive dorsal roots are anaesthetised, no sensory loss is apparent. The dermatomes for pain and temperature are very similar, but those of touch tend to be more extensive. It would be more appropriate if textbook diagrams did not show dermatomes between lines but between thick smudges!

330 Tuberculosis of a vertebral body may indeed be painful as a result of bony involvement but one would at first expect that pathology of T9 vertebral body would give thoracic pain posteriorly. The patient's acute abdominal pain is explained by the fact that the pain is referred to the longest fibres of the spinal nerves involved. The T9 spinal nerve has its anterior cutaneous branches in the skin between the xiphoid process and the umbilicus. It is here that referred pain from the back may be felt.

331 Shingles is the term used to describe a herpes zoster infection. This tends to pick a single nerve, often a mixed spinal nerve or a branch of the trigeminal nerve in the face, and the classic sign of this disease is a vesicular rash along the distribution of that nerve. In this way it maps out the dermatomes. The skin around the umbilicus is supplied by T10 spinal nerve and, more specifically, by its anterior cutaneous branches.

332 This narrowing of the aorta usually occurs in the vicinity of the ductus arteriosus, which may be still patent. In adult coarctation, the collateral circulation usually includes branches of the subclavian artery such as the internal thoracic, transverse cervical and transverse scapular as well as the intercostal arteries. Some vessels of the thoracic inlet such as the vertebral and anterior spinal arteries are also involved. After some time this collateral circulation, especially around the intercostal arteries, may be visible on a plain chest X-ray as notching of the inferior border of the ribs. This is usually seen in older children and adults, and is due to the erosion of the ribs by tortuous dilated intercostal arteries.

333 Interchondral joints in the lower thoracic region are prone to subluxation, especially following trauma. The eighth, ninth and tenth costal cartilages may be involved, and it is extremely easy to trap an intercostal nerve when the costal margin 'clicks' out of place. This may cause stimulation of the intercostal nerve and hence spontaneous contraction of its myotome. In the case of the ninth intercostal nerve this will cause contraction of the three sheet-like abdominal muscles, viz. external and internal obliques and transversus abdominis. It will also cause a segment of the rectus abdominis to contract just superior to the umbilicus. This clinical problem of interchondral subluxation is often misdiagnosed as cardiac or upper abdominal pathology, when careful examination will reveal a 'clicking rib' or 'clicking costal margin'.

334 The first rib is the shortest, widest, strongest and flattest as well as having the most acute curve of all the ribs. Its costochondral articulation lies just below and lateral to the sternoclavicular joint and, radiologically, looks as if it is broken off at this point, especially if seen in a posteroanterior view of thorax. The costochondral joint of the first rib is also often visible on an oblique view of the sternum.

335 Cervical ribs occur in about 0.5% of the population and are often bilateral. Though previously thought to be the cause of 'cervical rib syndrome' or 'scalenus anterior syndrome', it is now evident that the general term of 'thoracic inlet syndrome' is more appropriate (outlet in USA), some of these people having no cervical ribs. The essential symptom of these syndromes is involvement of the lower trunk of the brachial plexus. In those without a rib, this may be caused by the neurovascular bundle being trapped between the first rib and clavicle or by a fibrous band from C7 to the first rib. The symptoms include tingling, numbness and even pain in the distribution of the T1 dermatome, which lies along the medial border of the forearm. Vascular changes and wasting of the intrinsic muscles of the hand may also accompany this neurological deficit.

336 The body of the sternum is formed from four sternebrae which ossify from either single or bilateral centres, usually during the last few months

of intrauterine life. Fusion normally takes place from below upwards between the seventh and fifteenth years. So, if three separate sternebrae are visible, one would think of a child of that age. By 25 years the body is unified, and in some 10% of people the manubriosternal joint is fused after the age of 30.

B Information answers

337 The right lung has three lobes, the left one only two. The right lung is larger and heavier than the left lung but it is shorter and wider because the right dome of the diaphragm is higher and the heart and pericardium bulge more to the left. The anterior margin of the left lung has the deep cardiac notch.

338 The surface marking of the dome of the pleura is a curved line, convex upwards, drawn from the sternoclavicular joint to the junction of the medial and middle thirds of the clavicle. The highest point of the line is 2.5 cm above the clavicle.

339 The parietal pleura is supplied by somatic nerves and is therefore sensitive to pain. The costal part is supplied by the relevant intercostal nerves. The diaphragmatic part is supplied by the phrenic nerve centrally and the lowest intercostal nerves peripherally. The mediastinal portion is also supplied by the phrenic nerve.

   In contrast, the visceral pleura is supplied by autonomic fibres and is insensitive to pain.

340 The pulmonary ligament is situated at the root of the lung where the visceral and parietal layers of pleura meet as a sleeve surrounding the structures passing to and from the lung. This sleeve hangs down inferiorly as the pulmonary ligament. It is supposed to expand to allow movements of the lung root and expansion of the pulmonary veins.

341 The pleural cavity is the space between the parietal and visceral layers of pleura. It is normally collapsed and contains only a few millilitres of lubricant fluid but may become abnormally distended with air or fluid.

342 The lymph vessels travel along the bronchi to the hila of the lungs, filtering en route through many

lymph nodes within the lung substance. There are numerous lymph nodes at the lung root (tracheobronchial nodes). The lymph drains superiorly from these nodes as the bronchomediastinal trunk, one on each side of the body, which ascend adjacent to the trachea and end in the thoracic duct on the left and in the right lymphatic duct on the right.

343 The lungs do not completely fill the pleural cavity but leave recesses where two layers of parietal pleura are in apposition. The costodiaphragmatic recess is found posteriorly and laterally where the pleura covering the diaphragm is in contact with that covering the ribs. The costomediastinal recess is situated between the mediastinal pleura and that covering the adjacent left border of the sternum and ribs. Its presence is due to a deficiency of the left lung in this area, termed the cardiac notch.

344 Each main bronchus divides into lobar bronchi. These each divide again to form segmental bronchi. The part of each lobe supplied by one of these tertiary bronchi is a bronchopulmonary segment.

345 The middle lobe of the right lung is situated in the anteroinferior part of the right thoracic space between the oblique and transverse fissures. The oblique fissure runs in an oblique line from the spine of T3 to the sixth costochondral junction. The transverse fissure runs from the right fourth costochondral junction to the oblique fissure which it meets in the mid-axillary line.

The right middle lobe corresponds to the lingula of the upper lobe on the left.

346 The bronchial arteries are usually branches of the descending thoracic aorta. They vary in number – normally there is one on the right and there are two on the left. They anastomose to some extent with the pulmonary arteries.

Some of the bronchial venous blood passes to the pulmonary veins and thence to the left atrium. The remainder forms bronchial veins which drain, on the right, into the azygos vein and, on the left, to the superior intercostal or hemiazygos vein.

347 This appears in the right lung in about 1% of people, when the apical part of the apicoposterior

bronchus grows superiorly, medial to the arch of the azygos vein, instead of lateral to it. As a result, the azygos vein lies at the bottom of a deep fissure in the superior lobe (producing a linear marking on a chest X-ray that separates the lung apex from the remainder of the superior lobe).

B Applied answers

348 Pleurisy is inflammation of the pleurae but causes little discomfort if only the visceral pleura is involved because this is insensitive to pain and touch due to its autonomic innervation. However, the parietal pleura is supplied by the intercostal nerves (costal pleura), phrenic nerves (mediastinal and central diaphragmatic pleura) and the lower five or six intercostal nerves (peripheral diaphragmatic pleura). The regions supplied by the lower intercostal nerves may present as referred pain to the dermatomes of the upper abdomen and may therefore be indistinguishable from the pain of gastrointestinal origin.

349 The consequences obvious to the patient are often severe sharp pain, but in anatomical terms the important fact is that the pleural cavity is now opened and air can rush in, causing a pneumothorax with collapse of the lung involved.

350 A pleural effusion is a collection of excess fluid within the pleural cavity. Normally this contains only a thin film of pleural fluid, which is just enough to act as a lubricant for the opposing visceral and parietal layers. The exact position of this fluid will depend on gravity and the posture of the patient. In bed-ridden patients it tends to collect at the base of the lungs, especially posteriorly where it is easily detected by percussion.

351 A pneumothorax may be drained with the aid of a catheter and an underwater seal apparatus at one of two commonly chosen sites. These are the second intercostal space in the mid-clavicular line and the fifth or sixth intercostal space in the mid-axillary line. In the anterior approach the trocar and cannula pass through the anaesthetised skin, superficial fascia, pectoralis major and minor before penetrating the intercostal muscle fibres. Immediately deep to the thin membrane of

the innermost intercostal muscle is the parietal pleura, which acts as the external limit of a pneumothorax. The more lateral approach is preferred by patients because the pectoral muscles are not involved, thus giving pain-free movement of the arm.

352 The cervical pleura resembles a dome extending from the junction of the medial and middle thirds of the clavicle to the sternoclavicular joint. The highest point of the dome lies some 2.5 cm above the clavicle and the apex projects into the base of the neck. In this region, the pleural fascia is thickened to support the pleura and is known as the suprapleural membrane (Sibson's fascia). The veins used for catheterisation in the neck include the internal jugular, subclavian and brachiocephalic veins. The internal jugular vein drains blood from the brain and superficial parts of the face, and its surface marking is a band from the ear lobe to the medial end of the clavicle. The subclavian vein is a continuation of the axillary and begins at the outer border of the first rib. It unites with the internal jugular vein just behind the sternoclavicular joint, thus forming the brachiocephalic vein. All the above veins are immediate anterior relations of the cervical pleural domes, as many an inexperienced physician has found on performing a central venous catheterisation! Pneumothorax, haemothorax and even chylothorax are complications of this procedure, particularly after using the subclavian route.

353 It is prudent to avoid performing a left subclavian or brachiocephalic puncture due to the presence of the thoracic duct terminating in the angle between the internal jugular and subclavian veins. If, at the same time as damaging the thoracic duct near its termination, one lacerates the cervical pleura, a chylothorax can result.

354 The apicoposterior bronchus is found in the left upper lobe and is seen especially well on a left lateral or right anterior oblique bronchogram. There is considerable variation in bronchopulmonary segmentation; therefore bronchography is a less reliable method of correct identification than anatomical sectioning. However, the apicoposterior segmental bronchus quickly further

divides into the separate apical and posterior segments.

355 On bronchoscopy (insertion of a rigid tube via the mouth for examining the trachea and bronchi) a ridge called the carina (L. keel) is observed between the orifices of the main bronchi. Normally the carina is nearly in the sagittal plane and has a fairly definitive edge. If the tracheobronchial lymph nodes, in the angle between the main bronchi, enlarge (e.g. owing to the lymphogenous spread of cancer cells from a bronchogenic carcinoma), the carina may become distorted, widened posteriorly, and immobile.

356 The surface markings of the apical basal segment are fairly similar in both lungs, the superior limit of this bronchopulmonary segment being the oblique fissure posteriorly. The apex and upper border of this segment therefore lie along a line joining the spine of T3 and descend obliquely to follow the line of the sixth rib to the sixth costochondral junction. With the arm raised above the head, this lies along the medial border of the rotated scapula. This segment of lung tissue is a pyramid whose apex is along this line and whose base is about the area of the palm placed on the back just lateral to erector spinae. The apical basal segment is also called the superior basal or lower lobe segment.

357 The basic principle of effective postural drainage is positioning the patient so that secretions will drain from the diseased segment with the aid of gravity. The addition of a physiotherapist's skills has been likened to the attempts to extract the last drop of HP sauce from its original glass bottle by hitting the base whilst pointing the neck of the bottle downwards. Basal segments, therefore, may require the patient to be tipped head downwards and, in the specific case of a posterior basal segment, in a prone position.

358 An apical carcinoma often affects structures which lie in direct contact with the suprapleural membrane (Sibson's fascia). These include the sympathetic trunk, the cervicothoracic or stellate ganglion, and the roots of T1. Invasion of the sympathetic pathways of this level may cause a Horner's syndrome due to the interruption of the

sympathetic supply to the head and neck. This syndrome consists of a constricted pupil, flushed and dry skin of the face, drooping of the upper eyelid (ptosis) and retraction of the eyeball (enophthalmos). The drooped upper lid is due to partial paralysis of levator palpebrae superioris, whose superior tarsal portion of smooth muscle (Müller) is innervated by sympathetic fibres from the cervical chain. The pain in the little finger may be caused by direct involvement of the T1 nerve, whose dermatome lies along the medial border of the forearm and hand.

C Information answers

359 The heart doesn't rest on its base! The term derives from the somewhat conical shape of the heart, the base being opposite the apex. Its base (posterior aspect) is formed mainly by the left atrium, lying opposite T5–8 when supine (T6–9 when erect), and faces superoposteriorly towards the right shoulder. It is quadrilateral in shape; the aorta and pulmonary trunk emerge from, and the SVC enters, its most superior part.

360 No. The apex is formed by the left ventricle and points inferolaterally, posterior to the left fifth intercostal space in adults, 7–9 cm from the median plane, just medial to the left mid-clavicular line. The 'apex beat' is a clinical term, being the most laterally palpated part of the cardiac impulse, usually located by the tip of one finger just inferomedial to the left nipple in non-obese men and young women. In infants and young children it is more superior and lateral than the adult position. Body build determines its position in the adult.

361 The sternocostal (anterior) surface is mainly formed by the right ventricle and is visible in PA radiographs of the thorax. The diaphragmatic (inferior) surface is formed by both ventricles, but mainly the left one, as the posterior interventricular groove divides this surface so that the right ventricle occupies one-third and the left ventricle two-thirds. It is related to the central tendon of the diaphragm. The pulmonary (left) surface is formed mainly by the left ventricle and occupies the cardiac notch of the left lung.

362 The four borders of the heart are those of its sternocostal surface. The right border is the slightly convex right atrium (almost in line with the vena cavae), the nearly horizontal inferior border is mainly the right ventricle, the left border mainly the left ventricle (with a little of the left auricle), the superior border is formed by the right and left auricles with the infundibulum/conus arteriosus of the right ventricle between them.

363 The infundibulum (or conus arteriosus) of the right ventricle is its superior conical portion that gives rise to the pulmonary trunk. It forms part of the superior border of the heart.

364 These are all features of the internal surface of the right ventricle. The trabeculae carneae (L. little fleshy beams) are irregular muscular elevations, giving the internal wall a coarse sponge-like appearance. One of them crosses the cavity from the interventricular septum to the base of the anterior papillary muscle. This is the septomarginal trabecula (moderator band), carrying the right branch of the atrioventricular bundle. A thick muscular ridge, the supraventricular crest, arches towards and over the anterior cusp of the tricuspid valve, separating the ridged muscular wall of the right ventricle from the smooth walled conus arteriosus/infundibulum, and also separating the inflow (atrioventricular) orifice from the outflow (pulmonary) orifice.

365 In defiance of its name, the left atrium forms most of the base of the heart (posterior aspect). The interatrial septum slopes posteriorly to the right; therefore much of the left atrium lies posterior to the right atrium.

366 The valve of the truncus arteriosus of the embryonic heart has four cusps (anterior, posterior, right and left). This is split by the spiral septum to form two valves, each with three cusps. The heart undergoes a partial rotation to the left, hence they are named according to their embryological origin and not their anatomical position as posterior, right and left cusps for the aortic valve, and right, left and anterior cusps for the pulmonary valve (Nomina anatomica, 1989). Similarly, the aortic sinuses are also named right, left and posterior. This new

nomenclature also agrees with the names of the coronary arteries.

367 These are formed superior to each aortic valve cusp by dilation of the wall of the aorta; blood in them prevents the cusps from sticking to the wall of the artery and failing to close. The mouths of the right and left coronary arteries open into their respective aortic sinuses, but no artery arises from the posterior (non-coronary) sinus.

368 The interventricular septum is composed of a thin oval membranous part, and a thick muscular part. It slopes posteriorly to the right, and bulges to the right owing to the higher pressures in the left ventricle. Its margins correspond to the anterior and posterior interventricular grooves and their arteries on the surface of the heart. The membranous part, because of its separate and complex embryological origin, is a common site for a ventricular septal defect (VSD). Indeed, this defect accounts for 25% of all forms of congenital heart disease.

369 The aortic valve has three semilunar cusps, each having a fibrous nodule at the midpoint of its free edge. The thickened crescentic edge on each side of the nodule is the lunule (L. little moon). When the valve is closed, the nodules meet in the centre.

370 The cardiac skeleton, composed of fibrous and fibrocartilaginous tissue, forms the central support of the heart. Fibrous rings surround the atrioventricular canals and the origins of the aorta and pulmonary trunk. These rings give circular form and rigidity to these orifices, preventing the outlets to the ventricles from being dilated when the chambers of the heart contract and force blood through them. They also provide attachment for the valves. The cardiac skeleton, together with the membranous part of the interventricular septum, provide attachment for cardiac muscle fibres and form an electrical insulator between the atria and ventricles.

371 The right coronary artery typically supplies the right atrium and ventricle, the interatrial septum, the SA node (in 55%), the AV node (in 85%) and a variable part of the left heart. However, the clinically important SA nodal artery may arise from

the left coronary artery or its circumflex branch (45%). Running inferiorly in the coronary groove between the right auricle and ventricle, the right coronary artery gives off its right marginal branch at the inferior border of the heart, and on the inferior surface gives its largest branch, the posterior interventricular artery (opposite its origin, the AV nodal artery runs to the base of the IV septum to supply the AV node and bundle).

372 Variation of the coronary arteries and their branching pattern is common. In about 50% of cases the RCA is dominant (i.e. it crosses to the left side and supplies the LV wall); in 20% the LCA is dominant, and the pattern is balanced in 30%. Some people may only have one coronary artery, and in 4% an accessory coronary artery may exist!

373 The conducting tissue of the heart consists of specialised cardiac muscle. The sinoatrial (SA) node is the pacemaker. It is situated in the upper part of the sulcus terminalis superior to the opening of the superior vena cava into the right atrium. From here the impulse spreads over the atria to reach the atrioventricular (AV) node. This structure is found in the lower part of the interatrial septum, next to the septal cusp of the tricuspid valve. From here the impulse spreads down the atrioventricular bundle, which forms the only electrical and muscular continuity between the atria and ventricles. Where the interventricular septum becomes muscular the bundle divides into two – right and left. The right bundle runs down the right-hand side of the septum and crosses the right ventricle as the moderator band. The left bundle divides again and the two terminal branches run down the left side of the septum. All the bundles terminate in the ventricles as a subendocardial plexus of conducting tissue.

374 Most of the venous blood of the heart drains into the coronary sinus. This structure runs in the coronary sulcus on the posterior aspect of the heart and drains into the right atrium. It receives: first, the great cardiac vein which ascends in the anterior interventricular groove to reach the coronary sulcus; second, the middle cardiac vein which ascends in the posterior interventricular groove; and, third, the small cardiac vein which

accompanies the marginal branch of the right coronary artery along the inferior border of the heart. It also receives the oblique vein which lies behind the left atrium and the posterior ventricular veins which drain the diaphragmatic surface.

The rest of the venous blood either drains directly into the heart chambers via the venae cordis minimae (Thebesian veins) or reaches the right atrium directly via anterior cardiac veins which run across the front of the right ventricle.

375 The coronary arteries are thus named because they run in the coronary sulcus. This was originally in the coronal plane but has rotated during development to lie in the sagittal plane.

376 The left coronary artery arises from the left aortic sinus. It runs forwards for 1 cm between the pulmonary trunk and the left atrial appendage, where it divides into the anterior interventricular and circumflex branches. The former runs down the interventricular groove on the front of the heart. The latter continues in the coronary sulcus onto the posterior aspect of the heart, where it anastomoses with the right coronary artery.

There is much individual variation in the distribution of the coronary arteries. The left usually supplies the anterior part of the interventricular septum and adjacent portions of the right ventricle. It supplies the entire left ventricle, except for a small strip on its diaphragmatic surface. It also supplies the posterior aspect of the left atrium.

377 The margins of the cusps of the atrioventricular valves are connected by tendinous cords (chordae tendineae) to muscular projections of the apical part of the ventricular wall (papillary muscles). During ventricular systole the papillary muscles contract and, by pulling on the chordae tendineae, draw the cusps together. This prevents the cusps being forced into the atria during systole, thereby maintaining the competence of the valve.

378 The right atrium receives the superior vena cava, the inferior vena cava and the coronary sinus, which drain the upper parts of the body, the lower parts and the heart, respectively. It also receives its share of minute heart veins (venae cordis minimae) and one or two anterior cardiac veins.

379 There are three layers in the pericardium: an outer fibrous layer and two inner serous layers. The outer serous layer is the parietal pericardium which is closely adherent to the fibrous layer. The inner serous layer is the visceral pericardium (epicardium) which closely invests the heart. The pericardial cavity, containing a few millilitres of lubricant fluid, is found between the two serous layers.

380 The phrenic nerves are embedded in the fibrous pericardium. They innervate this structure with sensory fibres. They also supply the parietal serous pericardium in a similar fashion. The outer layers of the pericardium are therefore supplied by somatic nerves and are sensitive to pain.

381 The transverse sinus of the pericardium is found at the base of the pericardial cavity. It is a passage which exists between the reflections of pericardium around the arterial pole (aorta and pulmonary trunk) and venous pole (pulmonary veins and venae cavae). It is the remains of a hole which appears in the dorsal mesocardium during development.

The oblique sinus of the pericardium is found posterior to the heart. It is a recess shaped like three sides of a square, open inferiorly, and is formed by the reflection of the pericardium around the venous pole (pulmonary veins and venae cavae). This part of the pericardial cavity sits between the left atrium anteriorly and the oesophagus posteriorly.

382 The right and left phrenic nerves are embedded in the fibrous pericardium together with their accompanying blood vessels (pericardiophrenic artery and veins). The central tendon of the diaphragm and the fibrous pericardium are fused. The phrenic nerves supply both the pericardium and diaphragm.

C Applied answers

383 Occlusion may commonly occur in the anterior interventricular (anterior descending) artery, the right coronary artery, or the circumflex branch of the left coronary artery.

384 When there is an atrial septal defect, the AV bundle usually lies in the margin of the defect; its inadvertent destruction during surgery would cut the only physiological link between the atrial and ventricular musculature.

385 Drainage of a pericardial effusion or haemopericardium causing cardiac tamponade can be a life-saving procedure, and the casualty officer should be familiar with the various approaches. A wide-bore needle may be inserted through the left fifth or sixth intercostal space near the sternum, avoiding the internal thoracic artery which courses inferiorly a finger's breadth laterally to the sternum. This approach to the pericardial sac is possible because the cardiac notch in the left lung leaves part of this sac exposed. The needle is inclined superiorly and posteriorly at 45° to the patient's transverse plane and angled slightly medially to avoid the superior epigastric vessels running in the posterior rectus sheath. Having pierced the skin and superficial fascia, the needle passes through the anterior rectus sheath and left rectus abdominis muscle. The needle next encounters transversus abdominis, which forms the muscular posterior rectus sheath, and, after sliding over the superior surface of the diaphragm, passes through the fibrous and parietal serous pericardia. The tip of the needle will now lie between the parietal and visceral serous pericardium and hence in the cavity containing the effusion. A second approach is via the left part of the infrasternal angle, passing the needle superoposteriorly, thus avoiding the lungs and pleura.

386 This is an area of the left chest wall that gives a dull note to percussion. This area is devoid of overlying lung tissue, owing to the cardiac notch in the left lung which leaves part of the fibrous pericardium uncovered.

387 The transverse pericardial sinus is important to cardiac surgeons. After the pericardial sac has been opened, a finger and a ligature can be passed through this sinus between the great arteries and the pulmonary veins. By tightening the ligature the surgeon can stop the circulation through the great arteries while procedures are carried out on them.

388 The heart valves lie on an approximate diagonal line from the third left costosternal joint to the fifth or sixth right costosternal junction. Along this imaginary line across the body of the sternum they are arranged from superior to inferior in the following order – pulmonary, aortic, mitral and tricuspid (PAMT). To hear the sounds of these valves closing during the cardiac cycle, one listens to where the sound is transmitted along the vessel through which the blood flows after passing through the valve; for example, the aortic sound is heard best in the right second intercostal space anterior to the ascending aorta. The mitral valve is usually heard best in the region of the apex beat.

389 The right heart shadow is formed from above downwards by the right brachiocephalic vein, superior vena cava, right atrium and inferior vena cava. The left heart border is formed from the aortic knuckle, pulmonary trunk, auricular appendage of the left atrium and left ventricle.

390 The heart is an internal organ and hence its sensory innervation is experienced as referred pain to its dermatomes. Because the heart is innervated bilaterally, and the sensory input from the heart is via the upper four or five thoracic dorsal roots, the pain of myocardial infarction is usually of a severe, crushing, deep visceral nature in the T1–4 dermatomes and may also radiate along the medial side of the arm (normally left) and into the jaw and lower neck. Pain located in the region of the apex beat is normally of chest wall or supratentorial origin!

391 The foramen ovale is the passage through which fetal blood passes from the right to the left atrium. In about 75% of adults it is completely obliterated, i.e. the septum primum and septum secundum are fused. In 25% of individuals there is an overlap of the two septa, thus giving a physiologically normal person showing an apparently normal fossa ovalis; however, on detailed post-mortem examination there is no fusion and a probe or finger can pass through the still patent foramen ovale. Failure of any septal overlap will lead to a patent foramen ovale and an atrial septal defect (ASD).

392 The right ventricle is the anterior chamber of the heart which would be lacerated in this case. Remember that the right inferior border on a chest X-ray is the right atrium and the left atrium is the posterior chamber of the heart. The apex beat is where the right ventricle anteriorly and left ventricle posteriorly are felt as one pulsation on the chest wall. Classically, it is in the left fifth intercostal space along the mid-clavicular line, or about 9 cm from the midline.

393 Examination of the coronary arteries is nowadays a routine procedure, and, using selective arteriograms, the right and left coronary arteries can be seen individually. A selective right coronary arteriogram would normally reveal branches to the sinoatrial node as well as the atrioventricular node.

394 A lateral view of the lower oesophagus during a barium swallow will normally reveal three indentations of its anterior border: superiorly lies the arch of the aorta, then the indentation due to the left main bronchus and inferiorly the indentation due to the left atrium when enlarged. A patient with mitral valve disease may present with left atrial hypertrophy which can often be seen on the barium swallow.

395 The inner surfaces of both auricular appendages are rough and ridged by the musculi pectinati. The right atrium itself is divided by a ridge, the crista terminalis, anterior to which the atrial wall is rough and covered in musculi pectinati, whilst posteriorly the atrium is smooth. This is because the right atrium developed from the primitive sinus venosus (smooth) and primitive atrium (trabeculated musculi pectinati). Sometimes a faint groove can be seen on the exterior of the atrium, corresponding to the crista, and is termed the sulcus terminalis.

D Information answers

396 The mediastinum is arbitrarily divided into superior and inferior by a horizontal line through the sternal angle (T4). The inferior mediastinum is further subdivided into anterior, middle and posterior mediastina – the contents of the pericardium

forming the middle mediastinum – the anterior mediastinum being anterior to the pericardium and the posterior mediastinum posterior. The anterior mediastinum therefore lies below T4 and between sternum and pericardium. The anterior mediastinum contains the inferior part of the thymus gland, fat, some lymph nodes and the sternopericardial ligaments.

397 The thoracic duct commences in the abdomen at the superior end of the cisterna chyli. It enters the thorax in company with the aorta on the right side of that vessel to lie on the right of the oesophagus. It ascends the posterior mediastinum, passing behind the oesophagus and reaching its left side at T4. It runs superiorly in this position into the neck. It passes laterally behind the carotid sheath and then arches forwards over the dome of the pleura to drain into the confluence of the left subclavian and left internal jugular veins.

398 The splanchnic nerves arise from the lower ganglia of the thoracic sympathetic trunk: the greater splanchnic nerve from ganglia 5–9, the lesser splanchnic nerve from ganglia 10 and 11 and the least splanchnic nerve from ganglion 12. The greater and lesser nerves end in the coeliac ganglion on their respective sides. The least nerve runs behind the medial arcuate ligament and ends in the corresponding renal plexus.

These nerves contain preganglionic sympathetic fibres.

399 The transversely running structures in the mediastinum pass posterior to structures running vertically. Thus, for instance, the hemiazygos veins pass behind the descending aorta.

400 For the purpose of description, the oesophagus is divided into upper, middle and lower thirds.

The upper third is supplied by branches of the inferior thyroid arteries; the middle third by branches of the descending thoracic aorta; the lower third by branches of the left gastric artery.

The veins from the upper third run to the inferior thyroid veins, from the middle third to the azygos system and from the lower third to the left gastric vein, which is a tributary of the portal vein. Thus the venous blood from the upper two-thirds runs to systemic veins, and from the lower third to portal veins. The lower end of the oesophagus is therefore a site of portosystemic anastomosis.

401 The white rami communicantes comprise the total afferent supply to the sympathetic chains. They are found only between T1 and L2, running from the mixed spinal nerve to the sympathetic chain. They contain preganglionic sympathetic fibres whose cell bodies are in the lateral horns of the spinal cord. They are white because they are myelinated.

The grey rami communicantes are found throughout the spinal cord and comprise only part of the efferents of the sympathetic chain. They contain postganglionic sympathetic fibres and run from the sympathetic chain to the spinal nerve. Being devoid of a myelin sheath, they are grey.

402 The lowest cervical and the uppermost thoracic ganglion of the sympathetic chain are frequently fused to form the stellate ganglion. This is found anterior to the neck of the first rib and behind the origin of the vertebral artery.

403 The superior mediastinum is bounded by the manubrium anteriorly, the bodies of the first four thoracic vertebrae posteriorly and the upper parts of the two lungs laterally. It consists of loose connective tissue in which various structures are embedded. From anterior to posterior, these are: thymus gland, the great veins and the superior vena cava (of which they are tributaries), the aortic arch and its branches (the great arteries), the trachea, the oesophagus and thoracic duct.

Various nerves are found coursing through the superior mediastinum and, owing to the latter's asymmetry, they have different courses on the two sides. These nerves are the left and right phrenic nerves, the left and right vagus nerves, and the left recurrent laryngeal nerve. Lymph nodes are also to be found in the connective tissue.

404 The ligamentum arteriosum is a fibrous band which runs from the commencement of the left pulmonary artery to the lower concave surface of the aortic arch, and around which the left recurrent laryngeal nerve hooks. It is the remains of the ductus arteriosus of the fetus, which conveys blood

from the pulmonary trunk to the aorta, bypassing the lungs.

The ductus is derived from the left sixth aortic arch of early embryological development around which the recurrent laryngeal nerve winds, thus accounting for its position in the adult. On the right, the sixth and fifth arches disappear and the nerve is found looping around the fourth arch artery (right subclavian artery of the adult).

405 The left and right recurrent laryngeal nerves are branches of the left and right vagus nerves, respectively, and their courses are different on the two sides.

The left recurrent laryngeal nerve arises from the left vagus as it crosses the aortic arch. It hooks round the ligamentum arteriosum and ascends on the left in the groove between oesophagus and trachea.

The right recurrent laryngeal nerve arises from the right vagus as it crosses the first part of the subclavian artery. It loops underneath the artery, and ascends on the right between trachea and oesophagus in a similar fashion to the left nerve.

Both nerves pass deep to the lateral lobes of the thyroid gland, where they have an intimate relationship with the inferior thyroid artery, passing in front of it or behind it, or through its terminal branches. The nerves pass below the inferior constrictor muscle into the larynx to supply all the laryngeal muscles (except the cricothyroid) and sensation to the mucous membrane below the level of the cords.

406 As its name suggests, the left brachiocephalic vein drains blood from the head, neck and left upper limb via the left internal jugular and subclavian veins. Its tributaries include the thoracic duct, the left superior intercostal vein and the inferior thyroid vein. During its descent, it crosses the left common carotid artery, brachiocephalic trunk, the left vagus and phrenic nerves. It is separated from the manubrium sterni by the thymus (or its remnants) and by the origins of sternohyoid and sternothyroid muscles.

407 Common variations of the thoracic aorta include a right aortic arch, where the arch curves over the root of the right lung and passes inferiorly on the right side (sometimes passing behind the oesoph-

agus to reach its normal location), and a double aortic arch that forms a vascular ring around the oesophagus and trachea. This can cause dysphagia and surgical interruption of the ring is then required. Variations in the branching of the aorta include a retro-oesophageal right subclavian artery, again with possible dysphagia. 'Extra' vertebral arteries or less commonly, an accessory artery to the thyroid, the thyroidea ima, may be present.

408 Having already given off its cardiac branches high in the neck, the vagus nerve descends between the left common carotid and subclavian arteries in the thorax. When it reaches the left side of the aortic arch, it curves medially at its inferior border giving off the left recurrent laryngeal nerve (which hooks around the ligamentum arteriosum) while diverging posteriorly from the phrenic nerve (and separated from it by the left superior intercostal vein) to pass posterior to the left lung root, where it breaks up and contributes to the pulmonary plexus. Leaving as a single unmixed trunk, it joins fibres of the right vagus to contribute to the oesophageal plexus, leaving it as two mixed trunks that enter the abdominal cavity anterior (left trunk) and posterior (right trunk) to the oesophagus.

409 The following arteries are branches of the thoracic aorta: bronchial (only the two left ones), oesophageal (two to its middle third), pericardial, mediastinal, posterior intercostal (3rd–11th spaces), subcostal (supplying the anterior abdominal wall) and the superior phrenic (to the posterior surface of the diaphragm).

410 This is a variable, bilateral, asymmetrical, paravertebral venous system draining the back and the posterior walls of the thorax and abdomen. The azygos vein and its main tributary, the hemiazygos vein (L. *azygos* = unpaired) usually arise from the posterior aspect of the IVC and left renal vein, respectively. Together, they provide an alternative pathway of venous drainage from the abdomen and thorax. The azygos vein ascends in the posterior mediastinum, close to the right sides of the inferior eight thoracic vertebrae, and is covered anteriorly by the oesophagus as it passes first posterior to, then arches anterosuperior to, the right lung root to join the SVC. Here, it impresses the cadaveric lung. It communicates with the posterior inter-

costal veins, the vertebral venous plexi, and receives mediastinal, bronchial and oesophageal veins. The hemiazygos vein courses up the left side of the vertebrae, posterior to the thoracic aorta, as far as the level of T9, where it crosses to the right, posterior to the aorta, thoracic duct and oesophagus to join the azygos vein. A clinically important variation is when the azygos vein receives all the blood from the IVC, except that from the liver.

D Applied answers

411 Oesophageal atresia. Occurring in 1 in 2500 newborn babies, it results from incomplete division of the foregut into respiratory and digestive portions. In 90% of cases, the superior portion of the oesophagus ends as a blind pouch and a fistula connects the inferior portion of the oesophagus to the trachea.

412 Oblique X-rays during a barium swallow may reveal three normal thoracic oesophageal impressions caused by the aortic arch, the left main bronchus, and the right crus of the diaphragm. These impressions indicate where swallowed foreign objects are most likely to lodge, and where a stricture may develop after the accidental or suicidal drinking of a caustic liquid. Pathological compression may be observed from a dilated left atrium, aberrant aortic arches or a retro-oesophageal right subclavian artery.

413 Backflow of lymph from the thoracic duct can pass into the deep supraclavicular nodes, situated posterior to the lowest part of sternocleidomastoid muscle. Hence, this is probably why sentinel nodes (Virchow) are more commonly palpated on the left. This is known as Troisier's sign.

414 Because of the relatively short necks of children, their left brachiocephalic vein may ascend from the superior mediastinum, superior to the jugular notch, and enter the root of the neck. Its high location must be kept in mind when performing a tracheostomy in children.

415 The anterior mediastinum is located inferior to the manubriosternal angle, anterior to the pericardium and heart, and posterior to the sternum. Its major

component is the thymus gland, which also extends into the superior mediastinum. Normally, the thymus is relatively large in children and regresses with age but tumours of this organ may cause obstruction to the many structures passing through the superior mediastinum – in particular the superior vena cava and oesophagus, both of which are compressed relatively easily. A retrosternal goitre may also produce similar symptoms.

416 The fairly consistent diameters of the trachea and main bronchi are due to the strengthening cartilaginous bands and rings which maintain their shape even during inspiration. The trachea is a membranous tube supported by 15–20 U-shaped bands of hyaline cartilage which are deficient posteriorly. The lowest hyaline band in the trachea is similar in shape to a pair of bathing briefs. The carina separating the main bronchi may be likened to the crutch of the briefs.

417 The intrinsic muscles of the larynx are all innervated by the recurrent laryngeal nerves save the cricothyroid muscle, which is supplied by the external laryngeal nerve. The origins of these recurrent laryngeal nerves differ – the left leaving the vagus and hooking round the remnant sixth arch, the ductus arteriosus, whilst the right leaves the vagus in the root of the neck and hooks round the right subclavian artery (fourth arch derivative) just lateral to the first rib. An apical tumour may therefore readily affect the recurrent laryngeal nerves, especially on the right, by quite minimal local spread.

418 The lymphatic drainage of the oesophagus is through a plexus of vessels which pass into the posterior mediastinal nodes and along the left gastric vessels through the diaphragm to the coeliac nodes around the aorta. Because of the plexus, however, it is not rare for a lower oesophageal carcinoma to present as hard nodes palpable in the supraclavicular region due to the ascent of lymph into the mediastinum and neck.

419 Superiorly lies the indentation made by the arch of the aorta; then just inferiorly lies that of the left main bronchus whilst potentially the largest, most inferior impression is caused by the left atrium when enlarged.

420 Vomiting blood (haematemesis) in an alcoholic patient is commonly due to either a peptic ulcer or oesophagogastric varices. These varices are thin-walled, tortuous, dilated veins which lie in the submucosa and lamina propria and are formed at the site of portosystemic venous anastomoses when there is raised portal venous pressure. In this case the cause is likely to be cirrhosis. The submucosal veins receive tributaries from dilated subepithelial veins and drain to the veins found on the oesophageal serosal surface. These veins communicate with both the azygos vein (systemic) and the left gastric veins (portal).

421 Congenital tracheo-oesophageal abnormalities are relatively common quite simply because of the very close developmental origins of the two tubes. The laryngotracheal groove and tube differentiate from the floor of the foregut, so it is not surprising to find the oesophagus ending in a blind pouch whilst there is a fistula between the trachea and lower oesophagus. Sometimes the oesophagus opens into the trachea.

422 It is most likely to be inhaled into the right main bronchus, which is shorter, wider and nearly in the same direct line as the trachea. The right bronchus is only about 2.5 cm long compared with 5 cm on the left. Once inhaled, material such as peanuts, pins or simply gastric fluid will tend to pass into the right middle or lower lobes. However, in the unconscious patient lying on the right, inhaled material often favours the posterior segment of the upper lobe.

ABDOMEN

A Information answers

423 These are all fibrous structures associated with the rectus sheath. The linea alba, connecting the xiphisternum and symphysis pubis, is the strong midline fibrous fusion of the aponeuroses of the three anterior abdominal wall muscles. The linea semilunaris, marking the lateral border of the rectus sheath, curves upwards and convexly outwards bilaterally from the pubic tubercle to the costal margin at the ninth costal cartilage at the transpyloric plane. It marks the splitting of the

internal oblique aponeurosis to form the rectus sheath. Superiorly from a point just below the umbilicus, the external oblique aponeurosis passes anterior to the sheath, and that of the transversus posteriorly. Below this point all the aponeuroses pass anterior to the rectus abdominis muscle, forming an abrupt free posterior margin to the sheath, the arcuate line (Douglas).

424 The transpyloric plane (Addison) is a horizontal plane midway between the jugular notch of the manubrium sterni and pubic symphysis, roughly a hand's breadth below the xiphisternum, transecting the inferior border of the body of L1 vertebra. It is the key plane of the abdomen, passing through the pylorus of the cadaveric stomach (in life, it is actually a little lower than this, at about L2), the duodenojejunal junction, the neck of the pancreas with the portal vein forming posterior to it, where the fundus of the gall bladder touches the ninth costal cartilage and the linea semilunaris meets the costal margin, the origin of the superior mesenteric artery from the aorta, where the splenic vein courses behind the body of the pancreas and where the renal vessels course to and from the renal hila. The termination of the spinal cord is also just below L1.

425 The conjoint tendon is the fused inferomedial tendinous fibres of the transversus abdominis and inferior oblique muscles, originating from the inguinal ligament anterolateral to the deep ring, arching obliquely posteromedially as the roof of the canal, to pass behind the superficial ring thus reinforcing the weak posterior wall (transversalis fascia). Here it fuses with the linea alba and the pecten of the pubis. This double reinforcing role, much more pronounced at the superficial ring, is enhanced on contraction of the anterior wall muscles, when the tendon descends like a shutter, the roof of the canal coming tightly down over its contents, protecting the posterior wall from direct herniation.

426 Over most of the anterior abdominal wall, the superficial fascia consists of one layer with a variable amount of fat (up to several inches thick in the obese!). Just superior to the inguinal ligament, the superficial fascia divides into two layers, a fatty superficial layer (Camper's fascia) and a deeper

membranous layer (Scarpa's fascia), consisting of fibrous tissue and normally little fat. The superficial vessels and nerves course between these two layers. Camper's fascia is continuous with the superficial fascia of the thigh, while Scarpa's membranous layer is continuous with the deep fascia lata of the thigh and also with the superficial Colles' fascia of the superficial perineal pouch, which invests the penis and scrotum/labium majora. Scarpa's fascia loosely fuses with the deep fascia of the abdomen (a very thin, strong layer over the superficial muscles, which cannot be easily separated from them – it is thin to enable abdominal distension). Surgeons use Scarpa's fascia for holding sutures during closure of abdominal skin incisions. Between this layer and the deep fascia, there is a potential space into which extravasted urine may track superiorly from a ruptured spongy urethra in the superficial perineal pouch.

427 The main constituent of the inguinal canal is the spermatic cord in the male and the round ligament of the uterus in the female. The musculofascial coverings of the cord, developed from the layers of the inguinal canal as the testis descended, suspend the testis in the scrotum and contain the testicular 'services'. The ductus (vas) deferens is the readily palpable muscular tube conveying sperm from the epididymis to the ejaculatory ducts. Three arteries (all anastomosing posterior to the testis) contained within the cord are the testicular artery (an anterior branch of the aorta at L2 also supplying the epididymis), the slender artery of the ductus (from the inferior vesical artery) and the cremasteric artery (from the inferior epigastric artery, accompanying the cord to supply the cremaster muscle and other cord coverings). Up to 12 veins ascend from the testis, forming the pampiniform venous plexus (L. tendrils, the spiralling coiling organ of a climbing plant!) surrounding the ductus and accompanying arteries. The plexus can become dilated and tortuous, especially on the left where the testicular vein drains into the left renal vein (it drains to the IVC on the right) to form a varicocele (L. bag of worms). Lymph vessels drain the testis to the preaortic and lumbar nodes. Sympathetics running with the arteries convey afferent impulses which on testicular injury give excruciating deep abdominal visceral pain and a sickening sensation. The genital branch of the genitofemoral nerve (L2)

supplies cremaster within the cord, whereas outside it, the ilioinguinal nerve (the collateral of the iliohypogastric nerve, L1) gains access to the canal laterally between the muscular layers of the abdominal wall and not via the deep ring.

428 The gonads develop in the lumbar region (L2), deep to the transversalis fascia, between it and the peritoneum. The site of the inguinal canal in the fetus is first indicated by the gubernaculum, a ligament extending from the gonad through the anterior abdominal wall inserting onto the internal surface of the scrotum/labium majus. Later, a diverticulum of peritoneum, the processus vaginalis, follows the gubernaculum, and behind it evaginates the anterior abdominal wall to form the inguinal canal (not oblique at this early stage). In both sexes, the opening produced by the processus vaginalis in the external oblique aponeurosis becomes the superficial inguinal ring. The testes usually enter the canal in the posterior wall of the processus just before birth, with its stalk normally obliterating shortly after birth, leaving only that part surrounding the testis (tunica vaginalis). The scrotal ligament is the adult derivative of the gubernaculum. In contrast, the ovaries descend to a point just inferior to the pelvic brim, and do not normally enter the inguinal canal as the processus vaginalis has now obliterated. Part of the gubernaculum inferior to the ovary attaches to the lateral uterine wall, forming the 'ligament of the ovary' connecting ovary to uterus, continuous in the uterine wall with the 'round ligament of the uterus' connecting uterus to the labium majus (homologous to half of the scrotum). Persistent patency of the processus vaginalis in females, forming the 'canal of Nuck' (a clinical term), can give rise to an indirect inguinal hernia just as it can in young boys, but much less frequently. Cysts of the 'canal of Nuck' or labium majus may also develop from remnants of the processus vaginalis.

429 Small vestigial remnants of the embryonic genital ducts are rarely observed, unless pathological changes occur. The appendix testis (at its superior pole) and the appendix epididymis (related to its head) are vesicular remnants of the cranial ends of the paramesonephric and mesonephric ducts respectively.

430 The rectus abdominis muscle is formed by fusion of the ventral portions of the last six or seven thoracic myotomes to form a composite single muscle. These tendinous intersections are only of superficial thickness in the muscle.

431 The abdomen is divided up into nine regions by the construction of two lines in the transverse plane and two in the sagittal plane, rather as one would construct a frame for the game of noughts and crosses. The upper transverse line is the transpyloric plane at L1/2, which encircles the body at a point halfway between the suprasternal notch and the symphysis pubis. The lower transverse plane is the transtubercular plane at L5, drawn between the two tubercles on the iliac crests. The two vertical lines are the right and left lateral lines. They are constructed by raising perpendiculars to the transverse lines from each mid-inguinal point (halfway between pubic symphysis and anterior superior iliac spine). The median upper portion is termed the epigastric, with a right and left hypochondrial region on either side. The median part of the middle zone is the umbilical, with a right and left lumbar (or lateral) region on either side of it. The median lower portion is termed hypogastric or suprapubic, with a right and left iliac region to each side.

432 The testis is covered by a thick indistensible covering of fibrous tissue called the tunica albuginea. Each of the three erectile portions of the penis is covered by a tough fibrous fascia also termed the tunica albuginea. In the female the ovary possesses a tunica albuginea. It is a delicate layer of fibrous tissue situated immediately beneath the germinal epithelium.

433 The ductus (vas) deferens is easily palpated within the spermatic cord. It is supplied with blood from its own private artery, the artery to the ductus deferens. This arises from either the superior or the inferior vesical arteries, both being branches of the internal iliac artery.

434 The spermatozoa commence their journey at their site of production – the seminiferous tubules of the testis. These drain into the mediastinum of the testis where a network of anastomosing tubules is situated, the rete testis. From the rete

testis the spermatozoa pass along some 6–12 efferent ductules into the head of the epididymis. Here the ductules coalesce to form a single duct of the epididymis. This much-coiled tube is situated in the body and tail of the structure. From the tail of the epididymis the spermatozoa drain into the ductus deferens. This is the continuation of the duct of the epididymis and traverses the scrotum and inguinal canal. At the deep ring, it hooks round the lateral side of the inferior epigastric vessels and runs subperitoneally around the pelvic wall to the base of the bladder, looping anterior to the ureter as it does so. The distal portion of the ductus is dilated as the ampulla. The terminal end joins the duct of the seminal vesicle on the bladder base to form the ejaculatory duct, which pierces the prostate gland to enter the urethra at the side of the urethral crest.

435 The superficial fascia of the scrotum is entirely devoid of fat. It contains the dartos muscle (part of the panniculus carnosus, L. little fleshy cloth), which is a thin layer of unstriped fibres supplied by autonomic nerves, playing a role in temperature regulation of the testes by wrinkling the scrotal skin.

A Applied answers

436 The sites of the abdominal wall affected by herniae are the groin (inguinal and femoral herniae), umbilicus (umbilical hernia), and the rectus sheath (epigastric and Spigelian herniae).

An inguinal hernia may be 'indirect', where the peritoneal sac and its contents enter the canal through its deep ring, lateral to the palpable inferior epigastric artery, and follow the course of the ductus deferens sometimes down into the scrotum. This is often due to the presence of a small sac outpouching from the peritoneal cavity, a patent remnant of the processus vaginalis in baby boys. A 'direct' inguinal hernia pushes through the weak transversalis fascia of the posterior wall of the canal, medial to the inferior epigastric artery. This is often due to atrophy or partial paralysis of the action of the conjoint tendon, which closes down the canal roof like a shutter, protecting the weak posterior wall opposite the superficial ring. It is seen in older people, or in those with ilioinguinal nerve damage post appendicectomy.

A femoral hernia traverses the femoral canal (which is a fatty dead-space medial to the femoral vein in the femoral sheath). The medial sharp border of the lacunar ligament at the entrance to the canal (the femoral ring) often strangulates it, thus making this a surgical emergency. The neck of a femoral hernia is inferolateral to the pubic tubercle, whereas on reduction, the neck of an inguinal hernia is superomedial to it. An umbilical hernia may be due to a congenital failure of contraction of an umbilical scar (commoner in low birth weight and black infants), or the scar giving way in later life.

An epigastric hernia is due to linea alba weakness above the umbilicus, and is associated with obesity and an age greater than 40. A Spigelian hernia is due to weakness around the arcuate line of the rectus sheath, whereas a lumbar hernia is found in the lumbar triangle, posterior to the free border of the external oblique, above the iliac crest.

437 A gridiron (McBurney's) incision, performed for appendicectomy, involves each layer of anterior abdominal wall aponeurosis or muscle being split and retracted (but not cut) in the direction of the fibres of each layer, with resulting low incidence of incisional herniae. The ilioinguinal nerve (a collateral of the iliohypogastric, L1) may be damaged as it runs between the inner two layers of muscle if cutting is performed here, and the resulting weakness of the action of the conjoint tendon does nothing to improve the bikini line of young girls when they laugh or cough!

438 Abdominal incisions are carefully planned to gain adequate exposure and to yield the best cosmetic results, with their location depending on organs to be reached and the type of operation performed, with minimal disruption to the motor and sensory supply of the abdominal wall. Transverse incisions through the rectus abdominis, when performed away from the tendinous intersections, give minimum neurovascular disruption. Median incisions above or below the umbilicus through the linea alba are relatively bloodless and avoid major nerves. Paramedian incisions, 2 cm from the median plane, can extend from the costal margin to

the pubis. After the anterior layer of the rectus sheath is opened, the muscle is freed and laterally retracted, avoiding injury to its nerve supply. The posterior layer of the sheath and peritoneum are then incised to enter the peritoneal cavity. Suprapubic incisions are transverse, superior to the symphysis pubis, to gain access to the pelvic organs. Subcostal incisions, 2.5 cm inferior to the costal margin (thus avoiding the seventh intercostal nerve), give access to the gall bladder on the right and spleen on the left. The muscle-splitting (gridiron) incision for appendicectomy, commonly the oblique McBurney's incision (2.5 cm superomedial to the anterior superior iliac spine through McBurney's point), is now less popular than an almost transverse incision along the skin crease through the same point. In each case the external oblique muscle/aponeurosis is split obliquely in the direction of its fibres, followed by splitting of the internal oblique and transversus abdominis muscles, without their division. Carefully made, the entire exposure cuts no musculoaponeurotic fibres, provides good access, and avoids cutting, stretching and tearing the ilioinguinal and iliohypogastric nerves.

439 Because the anterior third of the scrotum receives afferent fibres from the L1 cord segment via the ilioinguinal and genitofemoral nerves in contrast to the posterior two-thirds which is mainly S3 via the perineal branch of the pudendal and posterior femoral cutaneous nerves, a spinal anaesthetic giving total scrotal anaesthesia must be as high as the first lumbar cord segment.

440 The skin of the anterior abdominal wall is divided into rough quadrants with respect to its lymphatic drainage. There is a fairly constant watershed at the level of the umbilicus, as if a belt were fastened here. Above the belt the lymphatic drainage of the skin goes to the anterior axillary or pectoral lymph nodes, while below it goes to the superficial inguinal nodes, and thus all quadrants will be affected in paraumbilical tattoo.

441 McBurney's point is two-thirds of the way along a line from the umbilicus to the right anterior superior iliac spine. It is a landmark for a classic gridiron incision used for appendicectomy.

442 Abdominal paracentesis is performed through various sites, the most popular being through the avascular linea alba, one- third of the way along a line from the umbilicus to the pubic symphysis, or over the iliac fossae lateral to the linea semilunaris. In the midline the trocar pierces in succession the skin, superficial fascia and the parietal peritoneum before entering the peritoneal cavity. In using a more lateral approach, the three flat abdominal muscles (external oblique, internal oblique and the transversus abdominis) are pierced prior to the transversalis fascia. It is normal clinical practice to empty the bladder prior to paracentesis to avoid its being punctured!

443 The segmental nerve supply to rectus abdominis is from the lower thoracic spinal nerves, which enter the muscle along its lateral border. During paramedian incisions the rectus muscle should, therefore, be retracted laterally to avoid trauma to these segmental nerves.

444 The Pfannenstiel incision is a transverse suprapubic incision performed for pelvic surgery. Depending on how far laterally it is extended, it involves the rectus sheath and muscles in the midline as well as the oblique and transverse abdominal muscles laterally. Due to the lateral innervation, it is rare to damage more than one segmental nerve, thus providing a strong postoperative scar. The closure is made in layers and, provided the anterior rectus sheath is well sutured, it is unnecessary to suture the muscle itself. The posterior rectus sheath at this level is almost non-existent, as the aponeuroses of the flat abdominal muscles lie anterior to rectus abdominis.

445 Collateral vessels, available in cases of inferior vena caval block, include the superficial and inferior epigastric veins, which carry the blood to the thoracoepigastric and superior epigastric veins and thence to the lateral veins and into the superior vena cava. Clinically, this passage of blood may be seen and felt as a tortuous varicosed large vein from groin to axilla. Other smaller channels occur between the pelvic veins, vertebral venous system and superior vena cava, as well as the azygos system.

446 If your patient was unlucky enough to have his ilioinguinal nerve trapped in the hernia repair, pain may be felt in the L1 dermatome, which includes the scrotum.

447 The testis is an internal organ and so the pain is felt as referred pain. Trauma to the testes is felt as a non-specific paraumbilical pain due to the testes' T10 innervation. Remember that it descends from the posterior abdominal wall. The T10 dermatome is in the region of the umbilicus.

448 The lymphatic drainage of the scrotum is similar to all that of the skin below the umbilicus and goes therefore to the superficial inguinal nodes. The testis, however, is an abdominal organ which descends during development from the posterior abdominal wall through the inguinal canal and into the scrotum. It brings its vascular supply from the abdominal aorta and its lymphatics drain to the para-aortic nodes situated just below the level of the renal veins. Thus patients with testicular tumours often develop para-aortic metastases, the inguinal nodes being free of disease.

449 A hydrocele is an abnormal collection of fluid between the parietal and visceral layers of the tunica vaginalis of the testis. It could be compared to a pleural effusion or ascites. Aspiration of this fluid with trocar and cannula is rarely of permanent relief, as the cause of increased serous fluid production has not been removed. It is therefore sometimes necessary surgically to obliterate the cavity of the tunica vaginalis to cure a hydrocele.

A varicocele is a swelling of the pampiniform plexus of veins draining the epididymis and testis. It is seen more commonly on the left side and may be accurately described as 'varicose veins in the scrotum'.

450 A vasectomy (deferential dochotomy, if the Nomina Anatomica, 1989, is strictly adhered to!) is usually performed in the upper part of the scrotum just before the spermatic cord enters the superficial inguinal ring. To reach the vas, the three major coverings will have to be incised, viz. the external spermatic fascia derived from the external oblique aponeurosis, the cremasteric fascia and muscle derived from the internal oblique muscle and the internal spermatic fascia from the transversalis

fascia. The easily palpable vas will now lie exposed and surrounded by numerous vessels and lymphatics.

451 The gubernaculum ovarii becomes both the ligament of the ovary and the round ligament of the uterus in the adult female. They are usually continuous and are attached to the uterus just below the uterine tube. The round ligament takes a course very similar to the male ductus deferens, passing along the inguinal canal and ending in the female scrotal homologue, the labium majus.

B Information answers

452 The stomach is found in the epigastrium and left hypochondrium, the spleen in the left hypochondrium. The bulk of the liver is in the right hypochondrium but it extends into the epigastrium and the left hypochondrium. The ileocaecal valve is in the right iliac region.

453 The spleen, liver and transverse colon are covered in peritoneum and are suspended from the abdominal wall by mesenteries. They are intraperitoneal. The kidney is found in the posterior abdominal wall behind the peritoneal cavity. Certain parts of the kidneys just happen to have peritoneum on their anterior aspect. They are extraperitoneal. The pancreas and duodenum are peculiar. As gut structures they originally developed intraperitoneally with mesenteries. However, these have become absorbed to leave the pancreas and duodenum stuck on the posterior abdominal wall covered with peritoneum only on their anterior aspect. These organs are described therefore as extraperitoneal, and, save for the first few centimetres, the duodenum is not easily movable.

454 This fat-laden peritoneal fold hangs down from the greater curvature of the stomach, connecting it to the transverse colon, spleen and diaphragm. It normally fuses during the fetal period, usually obliterating the inferior recess of the omental bursa, but these can be partially separated in adults, a procedure commonly performed as a surgical entrance to the omental bursa to gain access to the stomach bed. It contains a variable amount of extraperitoneal fat, being paper thin in

the emaciated and perhaps several inches thick in the obese! In life it usually reaches the pelvis (cadaveric contraction causes it to barely cover the intestine) before looping back on itself, overlying and absorbing the transverse colon, whose mesocolon (merged greater omentum and primitive dorsal mesentery) attaches it to the posterior abdominal wall along the inferior pancreatic border. It prevents the visceral peritoneum, covering the gut, from adhering to the parietal peritoneum, aiding bowel mobility. It also has a role as the 'policeman of the abdomen', moving towards and walling off inflamed or perforated viscera, bringing a blood supply from the right and left gastro-omental arteries.

455 A peritoneal ligament is a double layer of peritoneum that connects two organs, or an organ with the abdominal wall. It may contain blood vessels or their remnants, or indeed other viscera (e.g. the tail of the pancreas, an accessory spleen). The lesser omentum is described as three ligaments: the apron-like gastrocolic ligament inferior to, the gastrosplenic ligament on the left of, and the gastrophrenic ligament superior to the stomach. A peritoneal fold is a reflection of peritoneum with more or less sharp borders, often raised by blood vessels (e.g. the paraduodenal fold), ducts or obliterated fetal vessels (e.g. the median, medial and lateral umbilical folds). A peritoneal recess is a peritoneal fold forming a blind pouch (e.g. three at the duodenojejunal flexure, and at the iliocaecal area, the retrocaecal recess, a common home to the vermiform appendix).

456 A mesentery is any double-layered fold of peritoneum which attaches intestine to the abdominal wall. It provides the means whereby blood vessels, lymphatics and nerves supply the relevant portion of the intestine. The mesentery is a mesentery which attaches the jejunum and ileum to the posterior abdominal wall. It extends from the duodenojejunal junction (at the left side of L2), downwards and to the right for 15 cm, to terminate at the ileocaecal junction (over the right sacroiliac joint).

457 The omental foramen (foramen of Winslow, epiploic foramen) leads from the greater sac of the

peritoneum into the lesser sac. The foramen admits two fingers and its boundaries are:

anteriorly the free border of the lesser omentum with the three structures contained therein (hepatic artery, hepatic portal vein, common bile duct)

posteriorly the inferior vena cava and the right crus of the diaphragm

superiorly the caudate process of the liver

inferiorly the first part of the duodenum.

The admitting fingers are therefore between the two great veins.

Sometimes bowel herniates through the foramen and becomes incarcerated or strangulated. In these circumstances, decompression is difficult because all the boundaries of the foramen are vital structures and cannot be incised.

458 The omental bursa (lesser peritoneal sac) is a large peritoneal recess between the stomach with its attached lesser omentum (L. fat skin) and the posterior abdominal wall, giving considerable ease and freedom of movement to the stomach as it distends and empties. It is an extension of the peritoneal cavity into the invaginated right side of the dorsal mesentry of the stomach. Its inferior recess is between the duplicated layers of the gastrocolic ligament of the greater omentum, a potential space in adults. Its superior recess is limited superiorly by the diaphragm, posteriorly by the layers of the coronary ligament, anteriorly by the liver, with the inferior vena cava to the right and oesophagus to the left. The spleen with its associated ligaments represents the left limit of the omental bursa. Communication exists with the main peritoneal cavity (greater peritoneal sac) through the omental foramen (foramen of Winslow, epiploic foramen), located posterior to the free edge of the lesser omentum on the right side of the bursa. It is large enough to admit two fingers.

459 In development, the foregut has two mesenteries – ventral and dorsal. The remainder of the gut has only a dorsal mesentery. Thus the ventral foregut mesentery has a free inferior border which extends from the umbilicus to the foregut-midgut junction. The liver develops in this ventral mesentery dividing it into two: the falciform ligament from anterior abdominal wall to liver, and lesser omentum from liver to gut. In the adult, therefore,

the falciform ligament is a sickle-shaped double-layered fold of peritoneum. It extends in a line from the diaphragm and anterior abdominal wall as far inferiorly as the umbilicus. It runs onto the liver round which it then splits to enclose. It has a free lower edge in which is situated the ligamentum teres – the remains of the left umbilical vein.

460 The lesser omentum is a double-layered mesentery of peritoneum. It stretches in a line from the right side of the abdominal oesophagus, the lesser curve of the stomach and first 2 cm of the first part of the duodenum to the undersurface of the liver. The lesser omentum ends in a right free border, extending from the first part of the duodenum to the porta hepatis. In the free border are contained three vital structures: the hepatic portal vein posteriorly, the common bile duct anteriorly and on the right, and the hepatic artery anteriorly and on the left.

461 The liver is not wholly covered by peritoneum but has a bare area posteriorly, a consequence of the organ outgrowing its original complete covering of ventral mesentery. At the bare area, the liver and diaphragm are in direct apposition. The reflections of peritoneum from diaphragm to liver at the margins of the bare area constitute the coronary ligament. It has two leaves – an upper and a lower, at respective limits of the bare area. At the left and right margins of the bare area the leaves of the coronary ligament run for a short while in apposition before becoming continuous. These constitute the right and left triangular ligaments.

462 The lienorenal ligament is part of the dorsal mesentery of the foregut. It extends from the region of the left kidney as a two-layered fold of peritoneum to the spleen. It forms the left margin of the omental bursa. Between its leaves are the tail of the pancreas and the splenic vessels.

463 Normal females have two openings into their peritoneal cavity at the ovarian end of the Fallopian tubes. This communication (between peritoneal cavity and the exterior) may be readily visualised radiographically in a hysterosalpingogram. Contrast medium is introduced through the cervix, into the uterine cavity and along the tubes. A free

spill of medium into the peritoneal cavity demonstrates the patency of the tubes. This route, from the exterior to peritoneal cavity, may be followed by infection and is a documented, though rare, cause of peritonitis.

464 The median umbilical ligament is the fibrous remnant of the urachus. It is a midline fibrous cord joining the apex of the bladder to the umbilicus. The urachus originally joined the embryonic cloaca with the allantois in the umbilical cord. Sometimes it remains patent and a newborn baby will be found to have urine dribbling from its umbilicus. The medial umbilical ligaments are the fibrous remnants of the umbilical arteries of the fetus, originally branches of the internal iliac artery.

B Applied answers

465 Definite structures that form within the ventral mesentery or septum include the falciform ligament, liver and gall bladder and the lesser omentum. The other interesting structure from this region is the ventral part of the pancreas, which is an outgrowth from the primitive bile duct. It eventually becomes the lower part of the head of the pancreas and the proximal portion of the main pancreatic duct (Wirsung). This development explains the common arrangement of the openings of the pancreatic and biliary ducts at the duodenal ampulla (Vater). The ventral mesentery has a free caudal edge which becomes the anterior border of the omental foramen.

466 Ascites is an abnormal collection of fluid within the peritoneal cavity. It therefore lies between the parietal and visceral layers of peritoneum and is in a similar situation to a hydrocele or a pleural or pericardial effusion, each within their respective serous sacs.

467 The structures posterior to the lesser sac form the classic 'stomach bed' and include the diaphragm, pancreas, left kidney and suprarenal gland. Other structures include the transverse mesocolon, duodenum, spleen and its vessels.

468 The omental foramen is the opening into the lesser sac and marks the right or free edge of the lesser

omentum. This contains the hepatic artery, bile ducts and portal vein and the cystic artery, which usually arises from the right hepatic artery though it is prone to anomalies. Thus a bleeding gall bladder operation may be assisted by simple two-finger pressure on the hepatic artery lying in the free edge of the lesser omentum just anterior to the foramen.

469 For a volvulus to occur there must be a certain freedom of movement of the bowel concerned. The descending colon rarely has any mesentery worth discussing, whilst the sigmoid colon normally has a quite respectable mesocolon which, if too long and mobile, is likely to twist on itself. Interestingly, this condition of sigmoid volvulus is quite common in African races who have high fibre diet and relatively long sigmoid colons.

470 The appendix is a viscera, so early inflammation is evident as referred pain to the small gut dermatome of T10 (i.e. the region of the umbilicus). When, however, the local parietal (as distinct from visceral) peritoneum is involved, the pain is experienced at the site of inflammation, usually in the right iliac fossa near the classic site of McBurney's point.

471 The subphrenic spaces are potential spaces within the peritoneal cavity, which may become filled with pus and form subphrenic abscesses. The right and left subphrenic spaces lie between the diaphragm and the liver but are separated from each other by the falciform ligament. Both spaces have as their anterior boundary the abdominal wall and their superior limit the diaphragm. The posterior border on the right is the coronary ligament and on the left the left triangular ligament. Clinically, the right space is the more important.

C Information answers

472 The vermiform appendix is usually retrocaecal (64%) or pelvic (32%). Uncommonly it is retrocolic, a retroperitoneal location posterior to the ascending colon. Its base is fairly constant, usually deep to the surface marking of McBurney's point. The three taeniae coli of the caecum converge here to form a complete outer longitudinal

muscle coat for it, joining the caecum 2.5 cm inferior to the ileocolic junction. The appendix is relatively longer and narrower in infants and children. Although the human appendix is considered a vestigial organ, it has actually developed progressively in primates. During appendicectomy, the caecum is delivered into the surgical wound (not the transverse colon!) and the mesoappendix with its vessels is firmly ligated and divided. The base is tied, the appendix excised and the stump is sometimes cauterised and invaginated into the caecum. In unusual cases of malrotation or incomplete caecal rotation, the appendix may be found in the right upper quadrant due to a subhepatic caecum!

473 An ileal diverticulum (Meckel) is one of the most common malformations of the digestive tract. This blind-ending finger-like pouch is a remnant of the proximal part of the embryonic vitellointestinal duct (yolk stalk). The 'rule of twos' is a useful guide: it is two inches long, twice as common in males, located between two inches and two metres from the ileocaecal valve and occurs in 2% of the population! It projects from the antimesenteric border of the ileum, and may be connected to the umbilicus by a fibrous cord, or a fistula (which represents a persistent yolk stalk) with a 'raspberry tumour' at the umbilicus due to diverticular mucosa. It is of clinical significance because when inflamed it can mimic the symptoms of acute appendicitis. Also, bowel obstruction can occur from its fibrous cord, or even complete ileal volvulus using its cord as an axis of rotation! In addition, the wall of this ileal diverticulum may contain patches of gastric type epithelium which may secrete acid, bleed or ulcerate. Pancreatic tissue may also occur here, and be a rare site for an APUD tumour.

474 During their thoracic course, the vagi eventually come to lie on the oesophagus – the right vagus on its right and left vagus on the left. These nerves then divide up to form the oesophageal plexus. At the lower extremity of the oesophagus the vagi become reconstituted as right and left (mixed) trunks. However, owing to the rotation of the stomach during development, the reconstituted vagal trunks are situated anteriorly (left trunk) and posteriorly (right trunk) to the oesophagus, where

475 The stomach is innervated by autonomic nerves of the sympathetic and parasympathetic systems.

The sympathetic nerves are derived from segments T6, 7, 8 and 9. Their preganglionic fibres pass through the diaphragm as the splanchnic nerves and relay in the coeliac ganglia to be distributed to the stomach via blood vessels. The parasympathetic fibres are derived from the vagal trunks. Both vagal trunks run on their respective aspects of the lesser curvature and send branches to the anterior and posterior surfaces of the stomach. The anterior vagal trunk also sends branches to the liver via the lesser omentum. Recurrent branches from here sweep back to supply the pyloric region.

476 The stomach bed is composed of those structures which lie behind the stomach and are separated from it by the lesser sac (omental bursa). These structures are the diaphragm, left suprarenal gland (and upper part of left kidney), the pancreas, spleen, splenic vessels and the transverse mesocolon.

477 The artery of the foregut and foregut derivatives is the coeliac axis (trunk). The foregut extends as far inferiorly as the entry of the bile duct into the second part of the duodenum and consists of the lower oesophagus, stomach and the proximal duodenum, liver, spleen and proximal part of the pancreas.

478 The lymphatic drainage of the stomach closely follows its arterial supply and the nodes can therefore be conveniently divided into four groups. The lymph nodes around the left gastric vessels receive lymph from the left half of the lesser curvature. From this group, efferents pass to the coeliac nodes around the coeliac artery. The lymph nodes around the right gastric vessels receive lymph from the right half of the lesser curvature. From here, efferents pass along the hepatic artery and to the coeliac nodes. The left portion of the greater curvature sends its lymph along the short gastric and left gastro-omental arteries to nodes in the hilum of the spleen. From here, efferents pass to the pancreaticosplenic nodes along the course of the

(they pierce the diaphragm with the oesophagus and enter the abdomen.)

splenic artery to the coeliac nodes. Lymph from the right portion of the greater curve follows the gastro-omental artery, which has nodes along its length. Efferents follow the gastroduodenal artery to the coeliac nodes.

479 The small intestine changes its character gradually throughout its length, jejunum merging imperceptibly into ileum. However, various features distinguish proximal jejunum from terminal ileum. The jejunum has a thicker wall owing to the thicker and more numerous mucosal folds (valvulae conniventes) in this region. The mesentery has more fat between its layers in the ileum as compared with the jejunum. In the latter it is relatively transparent, forming 'windows' between the blood vessels. In the former, the fat makes it opaque. The mesenteric blood vessels form only one or two (anastomosing) loops or arcades to the jejunum. The ileum possesses up to half a dozen shorter layers of arcades. The jejunum is of greater diameter than the ileum. Furthermore, the ileum contains aggregations of lymphoid tissue (Peyer's patches), palpable on its antimesenteric border, which are absent from the jejunum.

480 The autonomic supply to the intestine may be divided into two: the sympathetic and the parasympathetic innervation.

The sympathetic innervation is derived from the thoracic sympathetic trunk as the splanchnic nerves. These contain preganglionic fibres and pierce the diaphragm to synapse in the coeliac plexuses situated on each side of the coeliac artery.

The parasympathetic supply is derived from the vagus. A large branch passes from the posterior vagal trunk in company with the left gastric artery, to the coeliac plexus. The parasympathetic fibres eventually synapse in their target organs, the postganglionic fibres being extremely short. From the coeliac plexus, branches of both sympathetic and parasympathetic systems pass with the coeliac and superior mesenteric arteries to the intestine.

481 The large bowel is usually of larger calibre than the small intestine. Its mucosa is featureless whereas the small intestine is folded and feels velvety, owing to the presence of villi. The colon carries fat-filled

tags in its peritoneal covering – the appendices epiploicae. The outer longitudinal muscle coat of the large intestine is not as complete as that of the small intestine. Instead it is condensed into three flattened bands, called taeniae coli, set at regular intervals around the circumference of the bowel. These taeniae are shorter than the rest of the large bowel and therefore throw it into characteristic sacculations. In contrast, the small bowel is of uniform width with no sacculations.

482 The junction between the midgut and the hindgut is situated between the right two-thirds and left third of the transverse colon. The midgut portion is supplied by the superior mesenteric artery via the middle colic branch, the hindgut portion is supplied by the inferior mesenteric artery via the left colic branch. These arteries anastomose along the mesenteric border of the colon by means of the marginal artery.

483 The pelvic mesocolon is a mesentery which attaches the sigmoid colon to the posterior pelvic wall. It has the shape of an inverted V. The apex is situated at the point of division of the left common iliac artery. The left ureter runs behind it. The right limb passes downwards and to the right and ends in the midline at the level of the third sacral vertebra. The left limb descends on the medial side of the left psoas muscle.

C Applied answers

484 Congenital hypertrophic pyloric stenosis is a thickening of the pyloric sphincter, found more commonly in first-born male children, which interferes with normal gastric emptying. Classically, it presents with the advent of projectile vomiting when the child is about 6 weeks old. It is cured by simple splitting of the hypertrophic muscle; it has been noted that the intramural ganglia cells are reduced in this region, though the aetiology of the disease is not fully understood.

485 Vagotomy is performed to reduce the gastric secretion, most commonly in cases of peptic ulceration. The anterior and posterior vagal trunks enter the abdomen through the oesophageal hiatus in the right crus of the diaphragm. The surgeon who

wishes to perform total or truncal vagotomy will find the oesophagus the most useful landmark. The anterior vagal trunk lies very close to the stomach and supplies the cardia, lesser curvature, liver and a clinically important branch to the upper border of pylorus and antrum (nerve of Latarjet). The posterior vagal trunk supplies the body of the stomach but its largest branch is via the coeliac ganglion to the rest of the fore- and midgut. As the vagus is both motor and secretory to the stomach, it is often necessary to combine vagotomy with a drainage procedure such as widening the pylorus (pyloroplasty). Due to its far-reaching innervation (L. *vagus* = a wanderer), a vagotomy is occasionally accompanied by severe gastrointestinal upset, most commonly seen as severe diarrhoea.

486 Occasionally, direct involvement of the pancreas – which is a posterior relation of the stomach and part of its 'bed' – may block the pancreatic ducts. More commonly, pancreatic problems are due to indirect lymph node involvement. The inferior part of the greater curvature of the stomach drains along the right gastro-omental vessels to the subpyloric group of nodes which lie in the head of the pancreas. It is via these nodes that the pancreas may be involved.

487 Following rupture of the bowel, gas may escape into the cavity. Clinically, this is most commonly seen in perforated peptic ulcer. A common cause nowadays for gas being in the cavity is following gas insufflation, mainly carbon dioxide, as used in laparoscopy, peritoneoscopy, hysterosalpingography or other X-ray procedures. It is demonstrated radiologically by seeing a thin gas shadow between the liver and diaphragm on a plain erect abdominal X-ray. This must not be confused with a normal stomach gas bubble.

488 On barium X-rays, the small bowel is seen centrally placed with plicae circulares (valvulae conniventes), commencing in the second part of the duodenum but disappearing about mid-ileum. These are especially well seen on a small bowel enema, double contrast X-ray. Because of this feature, the bowel has a 'feather-painted' appearance of barium. The large bowel, however, is positioned more peripherally, the caecum,

hepatic flexure and splenic flexure being obvious landmarks. On a standard barium enema sacculations can be seen especially in the transverse colon but this normal haustral pattern is best demonstrated by double contrast X-rays where both barium and air are used. This pattern is due to the relative shortness of the longitudinal muscle of the colon, forming three bands of taeniae coli, compared to that of the circular muscle fibres. This results in a bunching-up of the bowel wall into sacculations rather like material where a thread has accidentally been pulled.

489 Occasionally during fetal life the caecum will adhere to the visceral liver surface; more commonly however, following rotation of the mid-gut and its return to the peritoneal cavity, the ascending colon fails to develop, leaving the caecum in the right hypochondrium. This may result in appendicitis being mistaken for cholecystitis.

490 The region of the large bowel around the splenic flexure is that most commonly involved in ischaemic problems, as it is here that the middle colic artery of the superior mesenteric and the upper left colic artery of the inferior mesenteric artery anastomose. This region of the bowel is therefore supplied only by small terminal branches of the above and the very thin marginal artery which is easily blocked by atheroma. However, this area of relatively bloodless mesentery provides good emergency access to the lesser sac for splenic trauma, etc.

491 The sites of diverticulosis, or herniation of the mucosa through the muscle wall, are where the bowel wall is weakest. These sites are between the taeniae coli where there is only circular muscle fibres and particularly occurs at the sites where the blood vessels pierce these fibres.

492 The basic principle underlying colonic mobilisation is the partial reconstruction of its primitive mesentery. An incision is made laterally along the attachment of the visceral to the parietal peritoneum, in the paracolic gutter. The colon is then reflected medially by cleavage of the fused layers of fascia. During mobilisation, neither the vessels supplying it nor the ureter or kidney are disturbed. The four vessels lie anterior to the separated layers.

493 The most mobile parts of the colon are used for colostomies. Thus the usual ones are caecostomy, transverse colostomy and sigmoidostomy. When parts of the colon are removed, an adequate blood supply to the remaining segments must be preserved, otherwise necrosis will occur.

D Information answers

494 The pancreas (G. sweetbread) is derived from two foregut buds, dorsal and ventral, which grow into their respective mesenteries. The dorsal bud forms the neck, body and tail, whereas the ventral (bile duct) bud forms the head and uncinate process (L. hooked) of the pancreas. During development, the ventral bud somehow revolves 180° posteriorly around the duodenal circumference to lie in line with, and posterior to, the dorsal bud (the superior mesenteric artery being clasped between the two and thus appearing to pierce the pancreas). Usually, the primitive duct of the ventral bud forms the proximal end of the main pancreatic duct (Wirsung) and the distal part of the primitive duct of the dorsal bud forms the remainder, with a communication forming between the two primitive ducts. Frequently, the proximal end of the primitive duct of the dorsal bud also persists, forming an accessory pancreatic duct (Santorini) opening at the minor duodenal papilla (66%); here, some have a diminutive proximal section to their 'main' duct (20%) while others have a large one (10%). Often (44%) the accessory duct loses its connection with the duodenum. Fewer people (9%) have a main duct with no connection with the duct draining the head and uncinate process, the 'minor' duodenal papilla secreting the major part of the gland's digestive juices.

495 Branches from the splenic, gastroduodenal and superior mesenteric arteries supply the pancreas. Up to 10 small branches from the splenic artery supply the body and tail, as well as the large arteria pancreatica magna arising from it. The head is supplied by the anterior and posterior, superior and inferior pancreaticoduodenal arteries from the gastroduodenal and superior mesenteric arteries respectively. The pancreatic venous drainage is mostly into the splenic vein, with some into the portal and superior mesenteric veins.

496 Migration of the ventral bud behind the duodenum causes the ventral bud to lie behind the superior mesenteric vessels. In the adult pancreas the superior mesenteric vessels are therefore seen to pass between the dorsal and ventral buds (i.e. between the body and the uncinate process). It therefore appears as if the pancreas has been pierced by these vessels.

497 The pancreas lies on the posterior abdominal wall behind the lesser sac. The tail of the pancreas, however, turns forwards in the lienorenal ligament and usually lies in contact with the gastric impression of the spleen. It may be damaged during splenectomy.

498 The duodenum is some 25 cm long and is a C-shaped tube concave to the left. The head of the pancreas lies within this concavity. The duodenum is divided into four parts. The first part commences at the pylorus at the level of L1, 2 cm to the right of the midline, and runs upwards, backwards and to the right, anterior to and across the right side of L1. The second part of the duodenum runs vertically downwards on the right side of the vertebral column from L1 to L3. The third part runs horizontally to the left, anterior to and across the body of L3. The fourth part runs upwards and to the left, eventually to turn forwards at the level of L1, 2 cm to the left of the midline as the duodenojejunal junction.

499 Being a gut derivative, the venous drainage of the spleen must be to the portal system. The splenic vein commences by the joining of some half dozen splenic tributaries which leave the spleen at the hilum. The single vessel runs to the lienorenal ligament to the posterior abdominal wall, where it runs downwards and to the right. It is situated at a lower level than the artery and therefore runs behind the body of the pancreas where it receives the inferior mesenteric vein. It is not tortuous, unlike the artery. It ends behind the neck of the pancreas, where it joins the superior mesenteric vein to form the portal vein.

500 The spleen originally develops as separate discrete masses of splenic tissue, called splenunculi, in the dorsal mesentery of the foregut. Each splenunculus

has its own artery and vein. During development, the splenunculi fuse to form the definitive spleen. However, incomplete fusion at the anterior margin results in the characteristic notches. Sometimes fusion of the splenunculi is even less complete, resulting in accessory spleens. The typical division of the splenic vessels as they enter the spleen is due to the fact that the terminal vessels supply the remains of one splenunculus apiece. They are end vessels, with no anastomoses. This explains the predilection of the spleen to suffer typical wedge-shaped infarcts if any of these terminal branches are occluded.

501 The spleen is ovoid in shape with a convex diaphragmatic surface and a concave visceral surface, which contains the hilum centrally and has further concave impressions for stomach, colon and left kidney. The vessels enter and leave at the hilum. Its anterior border is characteristically notched. Its long axis lies along the left tenth rib, the spleen extending between the ninth and eleventh ribs. It extends usually no further anteriorly than the mid-axillary line. The numbers 1, 3, 5, 7, 9 and 11 summarise some important facts concerning the spleen (after H.A. Harris, a former Professor of Anatomy at Cambridge University, who died in 1968). It measures 1 x 3 x 5 inches, weighs 7 oz and lies between the ninth and eleventh ribs. This information does not yield itself easily to metric conversion!

502 Being a foregut derivative, the spleen must derive its blood supply from the foregut artery (coeliac artery). Thus the splenic artery arises from the coeliac artery and passes to the left behind the posterior peritoneal wall of the lesser sac. It is noted for its remarkable tortuosity and this causes it to alternately appear and disappear behind the upper border of the pancreas in this region. It crosses anterior to the left suprarenal gland and upper part of the left kidney and enters the lienorenal ligament to reach the spleen. On its arrival it divides into several terminal branches which enter the hilum. It supplies the pancreas as it passes along it via small pancreatic branches, and also the stomach via the left gastro-omental artery (which arises near its termination and runs in the gastrosplenic ligament) and the short gastric arteries.

D Applied answers

503 The main pancreatic duct usually joins the common bile duct to form the hepaticopancreatic ampulla (Vater), which empties via a common duct that pierces the posteromedial wall of the descending part of the duodenum at the major duodenal papilla. A gallstone passing along the extrahepatic biliary passages may lodge in the constricted distal end of the common duct with the possibility of bile backing up and entering the pancreatic duct, especially if the latter has a weak sphincter that is unable to handle the excess pressure. Swelling of the pancreatic head occludes the main duct with resulting inflammation of the body and tail. If the accessory duct connects with the main duct and opens into the duodenum, it may compensate.

504 Owing to the posterior relations of the pancreatic head, cysts or tumours in it may cause symptoms by pressing on the portal vein, perhaps causing ascites (an accumulation of peritoneal fluid) and on the bile duct, causing jaundice (the yellow/bronze colouration of the skin, mucous membranes and conjunctiva due to biliary retention). The pancreatic neck is posteroinferior to the pylorus of the stomach and tumours arising from the neck may obstruct the pylorus. The body rests on the inferior vena cava, and tumours here can obstruct this great vein causing swelling of the lower limbs. In unusual cases, the two primordia of the pancreas may grow abnormally and completely surround the descending part of the duodenum, forming a ring-like 'annular pancreas' that can produce duodenal obstruction during the perinatal period, or postnatally if inflammation or malignant disease develops. Carcinoma of the body often only presents late when the invasion of the posterior abdominal wall causes unremitting excruciating mid-back pain.

505 Severe blows to the left hypochondrium, especially when associated with rib fractures (the spleen is normally wholly under the cover of the ninth, tenth and eleventh ribs, mainly along the axis of the tenth) can tear the thin capsule and lacerate the soft, pulpy, friable parenchyma of the spleen. Splenectomy is often performed to arrest haemorrhage, although repair is preferable to removal in children

and adolescents because of the spleen's importance in immunity, especially against encapsulated bacteria. Left shoulder tip pain may be caused by phrenic nerve pain referral to the supraclavicular C4 dermatome from the central diaphragmatic peritoneum.

506 One or more accessory spleens may be found near the splenic hilum or embedded, partially or wholly, in the pancreatic tail in the lienorenal ligament, or between the layers of the gastrosplenic ligament. They are common (10%) and usually about 1 cm in diameter. Awareness of their possible presence is important because if not removed during splenectomy, the result may be persistence of the symptoms for which the splenectomy was performed (e.g. splenic anaemia).

507 The biliary duct system is best seen by either an intravenous cholangiogram or per- or postoperative cholangiograms, when water-soluble contrast medium is injected via the bile ducts through a tube left in situ during cholecystectomy. More recently, a percutaneous transhepatic cholangiogram has gained favour but this involves piercing the liver substance with a fine needle, and therefore presents more potential problems. An oral cholecystogram, though excellent for demonstrating the gall bladder, rarely shows the duct system. The pancreatic ducts may be outlined radiologically, the most popular approach being a retrograde cholangiopancreatogram using an endoscopic approach (ERCP). Quite often the distal pancreatic duct may be seen on a cholangiogram due to reflux from the duodenal ampulla.

508 The duodenum is a C-shaped section of bowel enclosing the head of the pancreas. As it leaves the pylorus, the first few centimetres are covered with peritoneum. However, for the remaining 20 cm, it is a retroperitoneal organ, making mobilisation difficult.

E Information answers

509 In 64% of cases the right hepatic artery crosses posteriorly to the biliary passages, and in 24% it crosses anteriorly. In 12% of cases an aberrant right

hepatic artery arises from the superior mesenteric artery. The right hepatic artery crosses anterior to the portal vein in 91% and posterior to it in 9% of cases.

510 The lobes of the liver are right, left, quadrate and caudate. The right and left lobes are separated by the attachment of the falciform ligament. The caudate and quadrate lobes are situated on the inferior (visceral) aspect. The quadrate lobe is situated anteriorly between the ligamentum teres, porta hepatis and gall bladder. The caudate lobe is situated posteriorly between the groove for the ligamentum venosum, the inferior vena cava and the porta hepatis. These lobes are demarcated by the attachment of peritoneum and are not functionally distinct. The quadrate and caudate lobes are functionally part of the left lobe and are supplied by the left branches of hepatic artery and portal vein, and are drained by the left hepatic duct.

511 The blood vessels supplying the liver are the hepatic artery and the portal vein. The hepatic artery is a branch of the coeliac axis, the foregut artery. This is logical, as the liver is derived from the foregut region. The hepatic artery supplies oxygenated arterial blood. The portal vein conveys blood to the liver from the gut (and gut derivatives). It is venous and rich in substances absorbed from the bowel. Both the hepatic artery and the portal vein reach the liver by ascending in the free edge of the lesser omentum. When they reach the liver they divide into right and left branches supplying each functional lobe. The liver is drained by interlobular hepatic veins which flow into three main veins (left, right and central), which in turn drain into the inferior vena cava as it passes through the liver – they are also important in supporting the liver.

512 The bare area of the liver is situated on the posterior aspect of the right lobe. It is so named because it is devoid of peritoneal covering, unlike the rest of the liver. The bare area is triangular in shape – base to the left, apex to the right. The apex is the right triangular ligament. The sides are the widely separated leaves of the coronary ligament. The base is the left margin of the inferior vena cava.

513 The surface marking of the liver may be projected on the body wall as follows. The upper border is a

horizontal line passing just below each nipple. The lower border is an oblique line running from the tip of the right tenth rib to the left nipple. Thus the bulk of the liver lies under the rib cage but is uncovered in the epigastrium. In a thin individual it may therefore be palpated, especially in deep inspiration, particularly if it is firmer than usual.

514 In the fetus, oxygenated blood full of nutrients returns from the placenta in the cord via the umbilical vein. The umbilical vein enters the umbilicus and runs in the free edge of the falciform ligament to the liver. Here it joins the left branch of the portal vein to form the ductus venosus. Some blood passes to the liver, but most bypasses it and runs to the inferior vena cava in the ductus venosus. (It is not necessary for the fetal liver to detoxify the nutrients in this blood, as this has already been done by the maternal liver.)

After birth, the cord is ligated and blood ceases to flow in the umbilical vein. Without the large quantity of blood flowing through it, the ductus venosus collapses. Eventually the ductus venosus and umbilical vein are obliterated and form fibrous cords – the ligamentum venosum and the ligamentum teres respectively.

515 The porta hepatis is where the blood vessels and ducts enter and leave the liver. It is situated on the inferior aspect of the organ. The upper extremity of the free edge of the lesser omentum is attached around it. The ducts and vessels bifurcate prior to their entry into the liver, so right and left branches of the hepatic artery and portal vein and right and left hepatic ducts are situated here. Autonomic nerves enter the liver at the porta hepatis, and lymph vessels emerge. Some lymph nodes are situated in this region.

516 The right and left hepatic ducts emerge from their respective lobes of the liver and unite to form a common hepatic duct. The common hepatic duct is some 3.5 cm long. It descends in the free edge of the lesser omentum. A diverticulum – the cystic duct – leaves its right side and runs to the gall bladder. The continuation of the biliary tree distal to this point is the common bile duct. The common bile duct descends in the right anterior aspect of the free edge of the lesser omentum, and then

passes posterior to the first part of the duodenum. It then lies in a groove in the posterior of the head of the pancreas. Usually it is joined by the main pancreatic duct, and together they open into the medial wall of the second part of the duodenum at the apex of the duodenal papilla (ampulla of Vater). The opening is guarded by a sphincter (Oddi). The gall bladder is a sac lying on the right inferior aspect of the liver. It is divided into a fundus, body and neck, the last being continuous with the cystic duct. The cystic duct runs down the lesser omentum, eventually to fuse with the common hepatic duct and form the common bile duct. It is bound by peritoneum to the undersurface of the liver. Its function is to concentrate and store bile and to contract and release bile into the duodenum to promote digestion of a fatty meal.

517 All the venous blood from the intestine and intestinal derivatives is gathered together in the hepatic portal vein for passage through the liver. In general, the foregut is drained by direct tributaries of the portal vein (mainly the splenic), the midgut by the superior mesenteric vein and hindgut by the inferior mesenteric vein. The portal vein is usually formed posterior to the neck of the pancreas by the confluence of the superior mesenteric and splenic veins. The inferior mesenteric vein usually drains into the splenic vein some 3 cm to the left of this point. The portal vein passes upwards and to the right, posterior to the first part of the duodenum, to enter the free edge of the lesser omentum, where it ascends to the liver.

518 There are certain areas in the body where the portal and systemic veins come into contact. Venous blood may then drain into either the portal vein or the caval system. Obviously these portosystemic anastomoses will be situated at the superior and inferior ends of the intestine where the gut meets body wall. There are other sites, however, where bowel is deficient of peritoneal covering and around the umbilicus.

Clinically, the most important anastomosis is at the lower part of the oesophagus where tributaries of the left gastric vein (portal system) communicate with oesophageal veins draining to the azygos system. There is a similar anastomosis at the anus where superior rectal veins (tributaries of the inferior mesenteric vein) anastomose with inferior rectal veins (tributaries of iliac veins). There are also anastomoses between the two systems behind the ascending and descending colon, pancreas and duodenum, and the bare area of the liver (i.e. organs which have lost their mesenteries and are in contact with the posterior abdominal wall). The paraumbilical veins run in the falciform ligament. They form a portosystemic anastomosis by connecting the left branch of the portal vein with superficial veins of the anterior abdominal wall.

E Applied answers

519 A common approach to liver biopsy is through the right eighth or ninth intercostal space in the mid-axillary line. Having passed through skin and superficial structures (including the intercostal muscles), the needle pierces the costal parietal pleura, crosses the costodiaphragmatic recess, through the diaphragmatic pleura and into the diaphragm. The needle now moves on respiration and the liver is biopsied in a quick 'one second' jab, the patient holding his breath in expiration. As both pleurae are normally pierced, a pneumothorax is an obvious potential hazard.

520 When the biopsy needle is in the liver substance, the diaphragm must remain static; otherwise the liver will move and be lacerated by the needle.

521 Riedel's lobe is a tongue-like extension of the right lobe of the liver and may even extend into the pelvis. It is thought to be a congenital abnormality, though tight lacing in women was once suggested as a causative factor! Its clinical significance is that although itself it is of no pathological significance, it may easily be mistaken for a tumour of the liver, omentum, pancreas or even a distended gall bladder.

522 The portal venous system is most easily seen by performing a splenoportogram X-ray. This can be obtained by percutaneous splenic puncture, operative mesenteric venography or via a selective splenic arteriogram. Most commonly, a percutaneous approach combined with splenic manometry is used in the investigation of portal hypertension. Following contrast injection, the vascular tree and especially the splenic, superior

mesenteric and portal veins are demonstrated, followed by a 'hepatogram' showing the intrahepatic venous tree. This technique may be used to see collaterals in cirrhosis, with oesophageal varices or to check the patency of a previous portacaval shunt.

523 A gallstone ileus usually happens due to previous inflammation of the gall bladder, causing it to adhere to its anterior relation, the duodenum. Later, the inflamed gall bladder, containing stones, ruptures or ulcerates into the duodenum and deposits its stony contents into the duodenum. A large stone may thus find itself obstructing the small bowel.

524 The pain of biliary disease may be localised to the right subcostal margin or quite commonly is felt as referred pain to the right shoulder or the inferior tip of the right scapula. The gall bladder receives branches from the anterior and posterior hepatic plexi. The afferent nerve supply is believed to pass via sympathetics through the splanchnic nerves and the right phrenic nerve. This phrenic connection is thought to explain the referred pain of the right shoulder in the region of the C4 dermatome.

525 A known history of gallstones and a sudden onset of jaundice makes one think that the stones have been expelled from the gall bladder, have passed along the cystic duct and into the common bile duct where they have caused an obstruction to the flow of bile. The normal width, as seen on X-ray, of the common bile duct is not greater than 12 mm, so a couple of medium-sized stones can easily obstruct it, especially at the sphincter (Oddi). Work on the detailed musculature of the sphincter has shown that there is sometimes both a pancreatic and a bile duct sphincter (Boyden) as well as the combined ampullary sphincter of Oddi.

526 Pain from biliary colic is referred to the right infrascapular, right upper quadrant and epigastric regions. This is because afferents from the cystic duct travel with the greater splanchnic nerve to T7 and T8 cord segments. Often, the inflamed gall bladder irritates peritoneum covering the peripheral part of the diaphragm, resulting in intercostal nerve afferents sharply localising the pain to the inferior thoracic wall. If more central diaphragmatic afferents are stimulated, the phrenic nerve

relays these mainly to the C4 segment of the cord, with referral of pain to the cutaneous distribution of the supraclavicular nerves over the right shoulder.

527 The cystic artery usually arises (76%) from the right hepatic artery in the angle between and posterior to the common hepatic and cystic ducts. Together with the liver, these two ducts form Calot's triangle. However, when it arises on the left of the biliary ducts (13% from right hepatic artery, 6% from the left hepatic artery, 3% from the gastroduodenal artery and 2% from the hepatic artery) it usually passes anterior to these ducts. Most errors in gall bladder surgery result from failure to appreciate common variations in the anatomy of the biliary system. Before dividing any structure, surgeons should clearly identify the three ducts (cystic, bile and hepatic) and the cystic and hepatic arteries. Before dividing the cystic duct and its artery, it is best to be certain that what appears to be the cystic artery is not a variation that also supplies the liver. Unexpected haemorrhage can also arise from an accessory cystic branch from the hepatic arteries. It can be controlled by compressing the hepatic artery in the anterior border of the omental foramen between index finger and thumb.

F Information answers

528 The right kidney is situated at a slightly lower level than the left and, owing to the asymmetry of the abdominal contents, the anterior relations on the two sides are different. The difference in height of the two organs accounts for minor differences in the posterior relations. The latter will, for the sake of ease of description, be ignored. The posterior relations of the kidneys are the diaphragm, the costodiaphragmatic recess of the pleura, the twelfth rib, psoas, quadratus lumborum and transversus abdominis. The subcostal nerve (T12) and iliohypogastric and ilioinguinal nerves (L1) are also posterior relations. The anterior relations of the right kidney (from top to bottom) are the right suprarenal gland, liver, right colic flexure and second part of duodenum, and small intestine. The anterior relations of the left kidney (from top to bottom) are the left suprarenal gland, stomach and

spleen, tail of the pancreas, left colic flexure and small intestine.

529 The kidneys are closely invested by a true fibrous capsule. Outside this capsule is a layer of fat – the perinephric fat. This fat is, in its turn, enclosed by the renal fascia which also surrounds the suprarenal gland. The kidney therefore is surrounded by three fascial layers: capsule proper, perinephric fat and renal fascia.

530 In development, the kidney is initially situated in the pelvis, the blood supply being derived from the nearest available artery – usually the common iliac artery. The kidneys ascend up the posterior abdominal wall, their blood supply migrating with them so that finally the renal arteries arise from the aorta at the level of L1. Sometimes the pelvic position may persist or the kidney may be arrested in its ascent. The renal artery in those instances would be derived either from the common iliac arteries or from the aorta near to its bifurcation.

531 Both renal veins drain into the inferior vena cava at the level of the transpyloric plane (L1). However, asymmetry occurs as a result of drainage of the gonadal veins. On the right, this vein usually drains directly into the inferior vena cava, below the level of the renal vein. On the left, the gonadal vein usually drains into the left renal vein.

532 From the minor calyces urine flows into one of two or three major calyces. These coalesce to form the pelvis of the kidney. Urine flows into the renal pelvis and then through a constriction – the pelviureteric junction – into the ureter. The ureter is 25 cm long and runs downwards vertically, blended with the peritoneum on psoas major, just medial to, and parallel with, the tips of the lumbar transverse processes. The ureter enters the pelvis by crossing the bifurcation of the common iliac artery, anterior to the sacroiliac joint, where a narrowing may be present as it passes over the pelvic brim. It then runs downwards and backwards to the region of the ischial spine, where it turns forwards and medially to enter the bladder. It traverses the bladder wall in an oblique manner, this intramural part being the narrowest point of the ureter. Urine is propelled down the ureter into the bladder by peristalsis. Calculi may become lodged at any of

the three narrow sites mentioned above, i.e. PUJ, crossing of pelvic brim and intravesical part. It should be said that the relations of the abdominal ureter are different on the two sides but the same in the two sexes, whereas the relations of the pelvic ureter are the same on the two sides but different in the two sexes.

533 The arterial supply to the ureters is derived from different vessels as the ureter descends. Being adherent to the peritoneum, vessels eventually reach the ureter via the peritoneum. Superiorly, in the abdomen, these vessels are derived from the renal artery, lower down from the gonadal artery. In the pelvis, blood is obtained from the superior vesical artery. Venous blood drains to corresponding veins.

534 The hilum of the kidney transmits – from anterior to posterior – the renal vein, the renal artery and the ureter.

535 The right suprarenal gland is pyramidal and is situated on the upper pole of the right kidney. Anterior to it is the right lobe of the liver, and its medial extremity lies behind the inferior vena cava. Posterior to it is the diaphragm. The left suprarenal gland is crescentic and is situated on the superomedial aspect of the left kidney, extending almost as far inferiorly as the hilum. Anterior to it are the pancreas and lesser sac and stomach. Posterior to it is the diaphragm.

536 A trick question for which we hope you didn't fall! There is no connection whatsoever between the sympathetic nervous system and the adrenal cortex, except perhaps for some vasomotor fibres supplying the blood vessels. The cortex is the site of steroid production. The connection is with the adrenal medulla. Preganglionic sympathetic fibres from the splanchnic nerves run to the medulla where they synapse. The cells of the medulla are modified postganglionic cells. They release their catecholamines into the circulation.

F Applied answers

537 Left-sided varicoceles are sometimes thought to be due to the fact that the loaded sigmoid colon lies

anterior to the left testicular vein. Partial obstruction of this vessel will lead to varicosities of the pampiniform plexus, a varicocele. Another anatomical explanation is that the left testicular vein drains into the left renal vein at an angle of about 90°, and any slight alteration in the angle of confluence may cause obstruction. However, haematuria with a palpable abdominal mass must make one suspicious of a renal mass. Classically, adenocarcinoma of the kidney spreads directly along the renal vein, and thus the tumour will obstruct testicular venous flow and cause haematuria and an enlarged left kidney.

538 Different methods are used to locate the exact position for renal biopsy, but most of them involve the use of X-rays and markers on the skin of the patient's back. The needle pierces in succession the skin, superficial fascia and the posterior layer of the lumbar fascia, and usually runs through the lateral edge of the quadratus lumborum muscle. It then passes through the anterior lumbar fascia and paranephric fat and is cautiously advanced until it lies in the renal fascia. Similar in method to a liver biopsy, the needle is advanced through the renal fascia and perinephric fat only when the patient holds his breath in inspiration. The operator must not hold the needle while the patient breathes; it must be allowed to swing freely, otherwise lacerations to the kidney may occur.

539 The surface markings of the kidney are most easily drawn on the back and overlie the erector spinae medially and quadratus lumborum laterally. Bearing in mind that they move about 2 cm inferiorly during inspiration, the upper pole lies opposite the T11 or 12 spine, the lower pole opposite the spine of L3. The hila of the kidneys are opposite the first lumbar vertebra. Due to the mass of the liver, the right kidney is normally situated slightly lower than the left, and the right lower pole may just be palpated in a thin subject on full inspiration.

540 The posterior relations of both kidneys include the diaphragm and the costodiaphragmatic recess of the pleural cavity. A biopsy involving any part but the lower poles might well pierce both layers of the pleura and cross the recess and thus expose the patient to a risk of pneumothorax.

541 A horseshoe kidney is a developmental abnormality caused by the fusion of the two metanephric masses and is found more commonly in men than in women. The normal kidney migrates cranially but the horseshoe kidney cannot because vessels such as the inferior mesenteric artery prevent such ascent. Often the single kidney receives its blood supply from the iliac artery, and, due to malrotation of the pelvis, ureteric obstruction is a frequent problem. Variants of this abnormality include a pelvic cake or central 'lump' kidney and the S-shaped kidney, both of which have anteriorly placed pelvis and anomalous vasculature.

542 Plain radiography may occasionally reveal the outline of the calyces and renal pelvis, particularly if a staghorn calculus is present. However, the more conventional techniques of intravenous pyelography (urography) and retrograde pyeleography give a much clearer picture. During a urogram it is normal to take a plain abdominal film prior to intravenous contrast injection to exclude or confirm radiopaque renal calculi. Following the injection, ureteric compression is often used to show better renal calyceal filling and films taken at intervals (e.g. 10, 15, 20, 30 minutes) demonstrate the passage of contrast from kidney to bladder.

543 As seen on a plain X-ray, the right ureter lies between the hilum of the kidney and the sacroiliac joint. It descends posterior to the parietal peritoneum on the psoas major muscle which separates it from the lower lumbar vertebral transverse processes. It is a vertical line down these processes which marks out the pathway of the ureter.

544 The common surgical approach to the kidney is retroperitoneally from the lumbar region, where the subcostal neurovascular bundle, iliohypogastric and ilioinguinal nerves and the ascending branch of the right and upper left colic arteries are vulnerable to injury, as they cross diagonally over the posterior surface of the kidney.

545 The ureter is supplied with pain afferents that run with the least splanchnic nerve, impulses entering the spinal cord at T12 and via the preaortic plexus into L1. This explains why the spasmodic, stabbing, agonising pain is referred to the lateral and inguinal regions of the abdomen. The pain moves from

loin to groin as the stone is forced down the ureter. The pain frequently reaches the scrotum or labia majorum.

546 Sometimes, the embryonic kidney fails to ascend into the abdomen and lies anterior to the sacrum. Although uncommon, awareness of this possibility of an ectopic pelvic kidney should prevent it being mistaken for a pelvic tumour and removed, as has happened. Also, it may cause obstruction or be injured during childbirth, as would a pelvic transplanted kidney. Pelvic kidneys do not receive their blood supply from the usual renal source, but from the inferior part of the aorta, iliac artery or median sacral artery.

547 There are three relatively narrow parts of the ureter at which calculi tend to lodge. These are the pelviureteric junction (PUJ), where the pelvis forms the descending abdominal ureter; the pelvic brim just anterior to the sacroiliac joint; and the ureteric orifice which traverses the bladder wall obliquely. This last is the narrowest site of all.

G Information answers

548 The lumbar plexus forms within the substance of the psoas major muscle. Emerging from its lateral border is the iliohypogastric nerve (L1), the lateral femoral cutaneous nerve (L2, 3 posterior divisions) and the femoral nerve (L2, 3, 4 posterior divisions). The medial border produces the obturator nerve (L2, 3, 4 anterior divisions) and the lumbosacral trunk (L3, 4), while from its anterior aspect emerges the genitofemoral nerve (L1, 2).

549 The aorta enters the abdomen by running behind the diaphragm at T12. It descends on the bodies of the lumbar vertebrae and terminates at L4 by bifurcating into the common iliac vessels. Its paired branches can be divided into those supplying viscera and those supplying the body wall. The visceral branches are three – the renal artery, gonadal artery and suprarenal artery. The body wall vessels are five – the inferior phrenic and four lumbar arteries. For the sake of completeness, it should be remembered that the aorta has three midline unpaired gut branches: the coeliac, superior mesenteric and inferior mesenteric arteries. It

also has an unpaired body wall vessel which arises at the aortic bifurcation – the median sacral artery.

550 The inferior vena cava commences in front of the body of L5, behind the right common iliac artery, by the union of the two common iliac veins. It ascends on the right side of the aorta, passes anterior to the right renal artery and runs in a groove on the posterior surface of the liver. It pierces the central tendon of the diaphragm to reach the thorax, and after a short intrathoracic course drains into the right atrium. It receives at its commencement the median sacral vein. As it ascends, it receives body wall tributaries symmetrically from each side – the inferior phrenic and two lumbar veins. It receives symmetrically the two renal veins at the level of L1 and three hepatic veins as it passes through the liver substance.

Because it is situated to the right of the midline, some of its tributaries are asymmetrical. Thus it receives the right gonadal vein and the right suprarenal vein directly. On the left, these vessels drain into the left renal vein.

551 The iliac fascia, which is thin above, thickens inferiorly on approaching the inguinal ligament, and is continuous anteriorly with the transversalis fascia (both of which contribute to the sheath of the femoral canal). The psoas fascia, attaching medially to the lumbar vertebrae and pelvic brim, is thickened superiorly to form the medial arcuate ligament of the diaphragm, and is fused laterally with the anterior layer of thoracolumbar fascia. Inferiorly, it blends with the iliac fascia. The dense quadratus lumborum fascia (continuous laterally with the anterior layer of thoracolumbar fascia) is attached to the anterior surfaces of the transverse processes of the lumbar vertebrae, the iliac crest and the twelfth rib. Superiorly, it is thickened to form the lateral arcuate ligament of the diaphragm and inferiorly, blends with the iliolumbar ligament. The thoracolumbar fascia, an extensive sheet covering the deep muscles of the back, has its lumbar part extending between the twelfth rib and iliac crest, laterally blending with the internal oblique and transversus abdominis muscles. It splits into three layers, with the quadratus lumborum between its anterior and middle layers, and the deep muscles of the back between its middle

and posterior layers. The thin anterior layer is attached, along with the psoas fascia, to the anterior surfaces of the lumbar transverse processes. Its dense posterior layer attaches to the spinous processes of the lumbar and sacral vertebrae and to their supraspinous ligaments.

552 The IVC is posteriorly related to the bodies of the lower three lumbar vertebrae, the right psoas major, the right sympathetic trunk, the right renal artery, the right suprarenal gland, the right coeliac ganglion and the right crus of the diaphragm. Anterior relations include the peritoneum, the superior mesenteric vessels in the root of the mesentery, the horizontal part of the duodenum and the head of the pancreas (with the portal vein and bile duct intervening). Superior to the first part of the duodenum, the IVC is posterior to the omental foramen; it then enters a groove on the inferior surface of the liver, between the right and caudate lobes to pierce the central tendon of the diaphragm at the level of T8. To the left of the IVC is the aorta, and to the right, the right ureter and kidney, and the descending part of the duodenum.

553 The tributaries of the IVC are the common iliac veins, third and fourth lumbar veins, the right gonadal vein, the renal veins, the azygos vein, the right suprarenal vein, the right inferior phrenic vein, and the three hepatic veins. The azygos vein connects the SVC and IVC, either directly or indirectly. It commonly arises from the posterior aspect of the IVC at the level of L2, with the renal veins, but may begin as a continuation of the right subcostal vein, or from the junction of that vein and the right ascending lumbar vein. The azygos vein enters the thorax through the aortic hiatus at level T12.

554 Three effective collateral routes are soon established on IVC obstruction to return venous blood from the lower body to the right side of the heart. Firstly, anastomosis between the superficial/inferior epigastric veins with the thoracoepigastric vein exists. Secondly, blood returns via tributaries of the IVC that anastomose with the vertebral system of veins, passing within the vertebral canal and bodies (this route also provides a way for malignant cells to spread to the vertebral bodies or brain from a pelvic tumour). Thirdly, blood may return via the

lateral thoracic veins, which connect the circumflex iliac with the axillary veins.

555 At one stage, during embryonic development, there are two inferior vena cavae. Usually the left one degenerates, but it may persist as a vessel of variable size which connects the left common iliac vein to the left renal vein. If large, it obscures the left sympathetic trunk.

556 The cisterna chyli is a frequently absent sac-like expansion of the inferior end of the thoracic duct, about 5 cm long and 6 mm wide. When present, it is located between the origin of the abdominal aorta and the azygos vein, to the right of the bodies of L1 and L2 vertebrae, posterior to the right crus of the diaphragm. It receives lymph from the right and left lumbar and the intestinal lymph trunks, as well as some descending efferents from the inferior intercostal lymph nodes. It is thus the final collecting chamber of lymph from the lower limbs, pelvis and abdomen. It is best seen on a first day (i.e. 24 hours after injection) abdominal lymph-angiogram.

557 The hypogastric plexus is a plexus of autonomic nerves. It is situated retroperitoneally in the lower abdomen and pelvis, anterior to the lower part and bifurcation of the abdominal aorta, and between the common iliac arteries – a right and left branch supplying each side. The part in relation to the aorta is sometimes called the superior hypogastric plexus, the part between the common iliac vessels, the inferior hypogastric plexus.

The sympathetic input to the plexus is from downward prolongations of the aortic plexus and branches from adjacent ganglia of the sympathetic chain. The parasympathetic input is from the pelvic splanchnic nerves (S2, 3, 4). The plexus supplies pelvic viscera with autonomic nerves by sending branches which run with blood vessels.

558 The lymph nodes of the posterior abdominal wall are distributed around the aorta. The para-aortic nodes lie to its right and left sides. The preaortic nodes receive the lymph from the intestine. The lymph from the gut follows the intestinal arteries to the aorta. The preaortic lymph nodes are therefore situated around the origins of the coeliac, superior mesenteric and inferior mesenteric arter-

ies. Efferent lymph from these nodes drains upwards as the intestinal trunk. The para-aortic nodes receive lymph from the kidneys, gonads and suprarenals. They also receive lymph from the deep layers of the abdominal walls and from their respective common iliac nodes. The lymph drains superiorly to the right and left lumbar trunks. The intestinal trunk and the lumbar trunks drain to the cisterna chyli, just below the aortic opening in the diaphragm.

559 The relationship of structures at the pelvic brim may be easily determined if one knows that structures concerning the body wall are posterior to structures concerning the genitourinary system and that, within the genitourinary system, genital structures are anterior to urinary structures. Thus at the pelvic brim the common iliac artery is the most posterior structure. It bifurcates as it crosses the pelvic brim into external and internal iliac arteries. The ureter crosses anterior to this bifurcation to reach the pelvis. The testicular vessels, having in their turn crossed the ureter anteriorly in the abdomen to gain its lateral side, cross the external iliac artery at the brim lateral to the ureter.

560 In order to answer questions concerning the relationships of structures in the posterior abdominal wall, it is necessary to realise that various organ systems bear a constant relationship to each other. Any structure concerning the gastrointestinal tract is anterior to any structure concerning the genitourinary tract, which in turn is anterior to any structure concerning the body wall. Furthermore, structures concerning the genital tract are anterior to structures concerning the urinary tract.

Thus psoas will be the most posterior structure on our list. The ureter runs down subperitoneally on psoas. During its descent on psoas, it will be crossed anteriorly by the testicular vessels running from medial to lateral. The left ureter and testicular vessels will have no relationship with the right colic artery. The right testicular vessels (and the ureter, which is posterior to them) will be crossed by the right colic artery, which is therefore the most anterior of all these structures.

561 The origin of the diaphragm may be divided into three parts: a sternal part, a costal part and a

vertebral part. These fibres arise circumferentially about the margin of the body wall and are inserted into a three-leaf-clover-shaped central tendon.

The sternal part consists of small slips which arise from the internal aspect of the xiphisternum. The costal part comprises slips which arise from the internal aspects of the lower six ribs, interdigitating with transversus abdominis.

The vertebral part consists of two crura – left and right. The right crus is the larger and arises from the right side of the bodies of L1–3. The left crus arises from the left side of the bodies of L1 and 2. The right crus is supposed to be the larger because the liver gives more resistance to diaphragmatic descent on that side. Posterolaterally, the medial and lateral arcuate ligaments, which are condensations over psoas and quadratus lumborum muscles, also act as diaphragmatic origins.

562 The median, medial and lateral arcuate ligaments are situated where the diaphragm is in contact with the posterior body wall. The median arcuate ligament is a fibrous connection between the medial borders of the two crura. The aorta runs behind this ligament. The medial arcuate ligament is situated lateral to the median arcuate ligament. It is the thickened upper margin of the fascia covering the anterior aspect of psoas. It extends from the upper part of the body of L2 to the transverse process of L1. Similarly, the lateral arcuate ligament is a thickening in the fascia covering the anterior aspect of quadratus lumborum. It is situated lateral to the medial arcuate ligament and extends from the tip of the transverse process of L1 to the lower aspect of the twelfth rib.

563 The central tendon of the diaphragm is formed from the septum transversum (which also forms the connective tissue of the liver) situated under the heart. The body wall sends a process inwards, one from each side, to meet the septum transversum. This is the pleuroperitoneal membrane. It eventually meets the septum transversum, with which it fuses. It then turns upwards as the pleuropericardial membrane, to separate the heart from the lung. This portion eventually becomes the fibrous pericardium. The fusion of the septum transversum and the pleuroperitoneal membranes leaves a gap posteriorly. This is filled by mesentery

cells from the oesophagus. The muscle of the diaphragm is derived from cervical somites which migrate to invade the connective tissue skeleton. This explains why the phrenic nerve (C3, 4, 5) supplies the diaphragm. The pleuroperitoneal membrane brings its own nerve supply with it as it migrates. Thus a rim of the diaphragm is supplied by sensory fibres from the lower six intercostal nerves.

564 There are three principal openings associated with the diaphragm: the caval and oesophageal openings through it and the aortic orifice behind it.

The caval opening is at the level of T8. It is in the central tendon of the diaphragm, whose stiffness acts to keep the vena cava open during all phases of the respiratory cycle. It transmits the inferior vena cava and terminal branches of the right phrenic nerve.

The oesophageal opening is at the level of T10, being 2.5 cm to the left of the midline. The oesophagus, vagus nerves, branches of the left gastric vessels and lymphatics are transmitted through this orifice. Its boundaries are muscular, as it runs through the left crus. However, a sling of fibres from the right crus cross the midline and loop round it, thus reinforcing the opening and creating a valve effect on inspiration, preventing gastric contents from being sucked up into the oesophagus.

The aortic orifice is at the level of T12 and is situated between the crura, which are bridged by the median arcuate ligament. It also transmits the thoracic duct and azygos vein.

In addition to the above openings, other structures pass directly through the diaphragm. Thus the splanchnic nerves (greater, lesser and least) pierce the crura. The sympathetic trunks pass behind the medial arcuate ligament on the psoas muscle. The subcostal nerve and vessels run behind the lateral arcuate ligament. The left phrenic nerve pierces the dome to the left of the midline, to supply the diaphragm from its abdominal surface. The superior epigastric vessels pass between the sternal and costal origins. Lastly the neurovascular bundles of T7–11 pass between the digitations of the diaphragm and the transversus abdominis to reach the abdominal wall.

G Applied answers

565 The iliopsoas muscle has extensive and clinically important relations to the kidney, ureter, caecum, appendix, sigmoid colon, pancreas, lumbar lymph nodes and the nerves of the posterior abdominal wall. When any of these structures is diseased, movements of this muscle may cause pain. As it lies alongside the vertebral column and crosses the sacroiliac joint, disease of these joints may cause spasm of iliopsoas, a protective reflex. An abscess resulting from tuberculosis of the lumbar spine (Pott's disease) tends to spread into the psoas sheath to produce a psoas abscess. Pus may track down its consequently thickened fascial tube, over the pelvic brim and deep to the inguinal ligament, causing swelling in the femoral triangle that can be mistaken for a hernia. Pus can also reach the psoas by tracking down from the posterior mediastinum when the thoracic vertebrae are diseased.

566 The coeliac ganglia are numerous nodular masses joined by variable connections to form the star-shaped coeliac plexus (solar plexus). It is the largest autonomic plexus and lies around the origin of the coeliac arterial trunk and on the crus of the diaphragm. The superior mesenteric ganglion lies at the inferior margin of the plexus. The coeliac plexus receives sympathetic contributions from the three splanchnic nerves of the thorax and via the first lumbar ganglia. Parasympathetic connections are mainly from the posterior vagal trunk and some smaller nerves from the anterior vagus. The vagus innervates the bowel as far as the splenic flexure. This immensely rich nervous input means that a direct blow to the solar plexus causes rapid changes in gut blood flow, leaving the victim with a feeling of impending doom and very 'winded'.

567 Tuberculosis of the spine (Pott's disease) is often seen as a paravertebral abscess in the thoracolumbar region. The pus may then track laterally under the fascia of the psoas muscle (psoas sheath) and, in doing so, may later appear as a fluctuant swelling below the inguinal ligament in the groin. The psoas abscess is easily mistaken for a femoral hernia.

568 The treatment of some patients with arterial disease in the lower limbs may include surgical

removal of two or more lumbar sympathetic ganglia with division of their rami communicantes, a procedure called a lumbar sympathectomy. Surgical access is commonly through the lateral extraperitoneal approach. The muscles of the anterior abdominal wall are split and the peritoneum moved anteromedially to expose the medial edge of the psoas major, along which the sympathetic trunk lies. On the left side it is overlapped slightly by the aorta (rarely by a persisting left IVC), while the IVC covers the right trunk – hence, a surgeon has to retract these structures medially, making these large vessels vulnerable. Their identification is not easy, as they lie in the groove between psoas major laterally and the lumbar vertebral bodies medially, obscured by fat and lymphatic tissue. Great care is taken not to inadvertently remove part of the genitofemoral nerve, lumbar lymphatic trunks, or ureter – uncommonly, pathologists see parts of these structures in the specimen pots sent to them!

569   The cisterna chyli is best seen on a first day (24 hours postinjection) abdominal lymphangiogram. It is found anterior to the bodies of L1 and 2, receiving vessels from the para-aortic glands and intestinal lymphatic trunks, and is thus the final collecting chamber of the lymphatics from the abdomen and lower limbs. It gives off the thoracic duct, which ascends posterior to the diaphragm and through the posterior mediastinum before opening into the left brachiocephalic vein.

570   The abdominal aorta lies in the midline and is often palpable in the epigastric region in the supine thin subject. Inferiorly, the aorta bifurcates to the left of the midline at the level of the iliac crests.

571   The testis drains lymph via the spermatic cord to the posterior abdominal wall from whence it developed. It normally drains to the para-aortic nodes lying between the level of the common iliac veins and renal veins.

572   The diaphragm is formed from four main parts: the septum transversum, which becomes the central tendon; the body wall tissues; the dorsal mesentery of the oesophagus; and the two pleuroperitoneal membranes, which close off the pleura

from the peritoneal cavity. Congenital herniae are most commonly seen in the posterolateral portion of the pleuroperitoneal canal (foramen of Morgagni). Defects in the left leaf of the diaphragm are much more common than in the right, and occasionally the whole central tendon fails to develop. Right leaf diaphragmatic abnormalities are often associated with other congenital defects. Oesophageal herniae are sometimes congenital but much more often are of a traumatic nature. These latter are divided into the sliding and rolling types, and are classically found in middle-aged, overweight women due to a weakness of the phreno-oesophageal ligament. Rolling herniae are also termed 'paraoesophageal', as the cardia remains in its anatomical position, but the fundus of the stomach protrudes into the thoracic cavity alongside the oesophagus. Most traumatic ruptures are left-sided, through the weak membranous vertebrocostal triangle. The stomach, bowel, mesentery and spleen may herniate, in that order of occurrence.

573 The diaphragm's motor supply is entirely from the two phrenic nerves (C3, 4, and 5) and its sensory supply is via both phrenic nerves and the lower intercostal nerves. The latter are only sensory to the peripheral portions of the diaphragm derived from the body wall. Central diaphragmatic pain is therefore referred to the dermatomes of the phrenic nerve and is felt over the upper shoulder region.

574 The inferior vena cava must pass through the central tendon because, during inspiration, the muscular portion of the diaphragm contracts and, should this compress the vena cava, then reduced venous return would result. In fact, the central tendon holds open the vena cava and reduced intrathoracic pressure causes the blood to be sucked towards the heart.

575 Because of the close relationship of the IVC to the inferior vertebral column, it and the common iliac vessels are vulnerable to injury during repair of a herniated nucleus pulposus of an intervertebral disc. A rongeur (strong biting forceps used for gouging and removing herniated discs) unintentionally pushed too far anteriorly is the culprit.

PELVIS AND PERINEUM

A Information answers

576 For a synovial joint, it is atypical on three counts. Firstly, fibrocartilage, not hyaline cartilage, covers the sacral surface of the joint. Secondly, the joint surfaces are jagged (although smooth in the newborn). Thirdly, there is very little movement. In the adult, the joint cavity is obliterated in places by fibrous bands which pass from one articular surface to the other.

577 The sacroiliac articulation depends entirely upon ligaments. The two joint surfaces lie in diverging planes – the weight of the fifth lumbar vertebra tends to push the sacrum down towards the symphysis pubis. There is no bony factor in its stability – the wedge-shaped sacrum is the reverse of a keystone. Opposing any simple gliding movement of the joint surfaces are the strong dorsal interosseous sacroiliac ligaments, and the iliolumbar ligament acting through the fifth lumbar vertebra. Opposing forward rotation of the sacral promontory are the sacrotuberous and sacrospinous ligaments. While these ligaments are intact, the bony surfaces so held in apposition are irregular enough to discourage gliding and rotation, but this bony factor is entirely dependent upon ligament integrity. Note that the bony surfaces are not weight-bearing: the body weight is suspended by the sacroiliac ligaments, which sling the sacrum below the iliac bones, hence the load tends to separate, not compress, the cartilage-covered articular surfaces!

578 Anterior to the alar of the sacrum cross the common iliac artery, the ureter, the apex of the sigmoid mesocolon, which receives the inferior mesenteric vessels, continuing as the superior rectal vessels, the lateral sacral vessels, the lumbosacral trunk (L4, 5) which courses down to the sacral plexus, and the sympathetic chain.

579 The pelvis is divided into two parts. The true pelvis lies below the pelvic brim, the false pelvis above. The pelvic brim is a line which can be followed round the inner circumference of an articulated pelvis and consists of the sacral prom-

ontory posteriorly, the iliopectineal lines laterally and the symphysis pubis anteriorly. The false pelvis forms part of the abdominal cavity. In the adult, the pelvic viscera are usually limited to the true pelvis.

580 The orientation of the pelvis in the anatomical position is with the pubic tubercles and anterior superior iliac spines in the same vertical plane. For convenience, a wall may make the ideal plane of reference.

581 The pelvic inlet is the line which demarcates the true pelvis from the false pelvis, and consists of the sacral promontory posteriorly, the iliopectineal lines laterally and the symphysis pubis anteriorly. It is an indistensible bony ring through which the fetal head must pass during labour. The pelvic outlet is bounded posteriorly by the coccyx, laterally by the ischial tuberosities and sacrotuberous ligaments, and anteriorly by the pubic arch. It is an incomplete osseoligamentous ring which is distensible and therefore its conjugates are not absolute. It may be stretched during labour.

582 The symphysis pubis is a secondary cartilaginous joint. The articular surfaces of the bones are covered in hyaline cartilage and are held together by a disc of fibrous tissue. A small amount of movement is possible at this joint. The sacroiliac joints are atypical synovial joints, involving the auricular surfaces of the ilium and the ala of the sacrum. Like all synovial joints, these articular surfaces are covered in cartilage and the joint is reinforced by anterior and posterior sacroiliac ligaments and the accessory sacrotuberous and sacrospinous ligaments. The posterior ligaments are particularly strong; together with the interlocking of the irregular articular joint surfaces, they prevent all but the smallest movement at these joints.

583 The inguinal ligament is the rolled-up free lower border of the external oblique aponeurosis and extends from the anterior superior iliac spine to the pubic tubercle. In the anatomical position, the superior pubic ramus lies behind and a little above the line of the ligament. The lacunar ligament extends from the pubic end of the inguinal ligament backwards and upwards to the superior pubic

ramus. Laterally, it has a free sharp crescentic margin which forms the medial limit of the femoral ring. It is on this sharp edge that femoral herniae frequently become strangulated.

584 The female pelvis is modified to allow passage of the fetal head through the pelvic inlet, canal and outlet during labour. In order to do this the diameters of the inlet and outlet are increased compared with the male. In the male, the greater sciatic notch subtends less than a right angle. In the female the coccyx is carried backwards by increasing this angle to 90°, thereby increasing the anteroposterior diameter of the pelvic outlet. Thus it can be said that the greater sciatic notch is L-shaped in women and J-shaped in men (L for Lucy, J for Johnny).

585 The differences in the bony pelvis between the two sexes arise because of two factors. First, the male is the stronger and the heavier. His pelvis is more strongly built, with more pronounced muscular markings. Second, the female pelvis is wider as a modification for the passage of the fetus in labour.

In the female, the pelvic inlet is oval to allow passage of the fetal head. In the male, it is heart-shaped owing to the forward projection of the sacral promontory. The pelvic canal in the female is short with almost parallel sides and no projections, to allow an easy passage of the fetal head. In the male it is long with tapering sides, and the ischial spines project into it. Thus in the female the subpubic angle is greater than 90°. In the male it is less than 90° and in the female the sciatic notch is L-shaped whereas in the male it is J-shaped. Furthermore, the inferior pubic rami are strong and everted in the male for the attachment of the crura penis; this marking is absent in the female.

A Applied answers

586 The bony pelvis can resist considerable trauma. Falls from a height onto the feet or buttocks may fracture the pubic rami and as a consequence rupture the urinary bladder. The femoral heads may be driven through the acetabulum into the pelvis, damaging other pelvic viscera. Persons under 17 years may fracture their acetabulum into its

three developmental parts, or tear their acetabular margins.

587 First of all, the ischial tuberosities are palpated, the woman usually lying in the lithotomy position, and an intradermal weal of local anaesthetic is produced halfway between the tuberosity and the rectum. With middle and index fingers in the vagina, the needle is guided just beyond and inferior to the palpable ischial spine, where local anaesthetic is injected. This blocks the main pudendal nerve as it passes dorsal to the spine prior to entering the pudendal canal (Alcock).

588 The true conjugate is the measurement from the sacral promontory to the upper and inner border of the pubic symphysis, whereas the diagonal conjugate is from the sacral promontory to the lower border of the symphysis.

589 The sacral hiatus is the bony landmark used for caudal epidural injections. Mainly used for back pain or during labour, epidural anaesthesia by this route is achieved by introducing the needle through the sacrococcygeal ligament, traversing the sacral hiatus, and injecting the local anaesthetic in the space occupied by sacral roots and filum terminale.

590 These 'dimples of Venus' are the posterior superior iliac spines, often seen in a well nourished woman as dimples but in thin individuals only as two bony prominences. They are a useful guide as to the approximate level of the end of the thecal sac as well as the S2 vertebra and the midpoint of the sacroiliac joint.

591 The subpubic angle is under the arch formed by the two ischiopubic rami and the pubic symphysis. In the male, this tends to be a sharp angle of about 60°. In the normal female gynaecoid type of pelvis, the angle is a more gentle arch of about 120–130°. Therefore, a woman with an acute subpubic angle has a more android pelvis and may indeed have problems during childbirth due to a reduced available outlet space.

592 The distance between the ischial spines is the diameter of the pelvic cavity. The normal measurement for interischial diameter is about

11 cm, so a patient presenting with an 8 cm diameter is a strong candidate for obstruction in the mid-cavity during labour. Depending on the size of her baby, this woman might require a Caesarean section.

593 Normal measurements of the brim or pelvic inlet are 11.5 cm for the transverse diameter and 12 cm for the diagonal diameter. The anteroposterior measurement is the true conjugate diameter, the transverse is the widest part of the brim and the diagonal or oblique measurement is from sacroiliac joint to iliopectineal eminence.

B Information answers

594 The superficial fascia of the anterior abdominal wall is divided into a superficial fatty layer (Camper's fascia) and a deeper membranous layer (Scarpa's fascia). Colles' fascia is the continuation of Scarpa's fascia in the perineum. It is attached anteriorly to the pubic arch and posteriorly to the free border of the perineal membrane. It sends prolongations into the penis and scrotum in the superficial fascia of these structures. Between Colles' fascia and the perineal membrane is the superficial perineal pouch.

595 The pudendal nerve is a branch of the sacral plexus. In company with the internal pudendal vessels, it leaves the pelvis through the greater sciatic foramen and turns immediately over the ischial spine into the lesser sciatic foramen to enter the pudendal canal (Alcock). The first branch, the inferior rectal nerve, is given off almost immediately. This nerve runs medially and breaks up to supply the external anal sphincter, perianal skin and part of levator ani. As the remainder of the pudendal nerve approaches the urogenital triangle, it divides into two – the dorsal nerve of the penis (or clitoris) and the perineal nerve. The former traverses the deep pouch and supplies the penis (or clitoris). The latter runs in the superficial pouch and supplies the muscles of the urogenital triangle and the skin of the scrotum (or labia).

596 The male urethra is divided into three parts: prostatic, membranous and penile (spongy). As its name suggests, the prostatic part traverses the

prostate gland. This part of the urethra lies in the pelvis. The membranous part passes through the urogenital diaphragm (the deep perineal pouch); that is, from superior to inferior, the superior fascia of the urogenital diaphragm, the sphincter urethrae muscle and the inferior fascia of the urogenital diaphragm (perineal membrane). The penile part runs through the corpus spongiosum of the penis. The membranous and penile parts lie in the perineum superficial to the levator ani muscle.

597 The deep perineal pouch (space) is a triangular-shaped space bounded superiorly by the superior fascia of the urogenital diaphragm and inferiorly by the inferior fascia of the urogenital diaphragm (perineal membrane). Anterolaterally these two fascial sheets are attached to the pubic arch and ischiopubic rami, which thus form the space's boundaries in this direction. Posteriorly, the space is bounded as the fasciae fuse along the free border of the urogenital diaphragm. In both sexes the pouch contains the membranous urethra and, surrounding it, the sphincter urethrae muscle, plus the deep transverse perineal muscles, the dorsal nerve of the penis (clitoris) and the blood vessels to the penis (clitoris). In the male it also contains the bulbourethral glands (Cowper). In the female the vagina passes through the deep pouch, piercing the sphincter muscles as it does so.

598 The superficial perineal pouch lies between Colles' fascia inferiorly and the inferior fascia of the urogenital triangle (perineal membrane) superiorly. Posteriorly these two structures fuse. Laterally its boundary is where these two fascial sheets attach to the ischiopubic rami but anteriorly it is in free communication with the sub-scarpal area of the anterior abdominal wall. The pouch contains the root of the penis (clitoris) and the associated muscles. These are the ischiocavernosus, bulbospongiosus and superficial transverse perineal muscles, which help maintain erection of the penis (clitoris).

599 The urogenital triangle is formed by the symphysis pubis anteriorly and the diverging ischiopubic rami laterally. Its posterior border is an imaginary line passing through the two ischial tuberosities.

600 In the anatomical position the penis is erect, so the dorsum is the part which faces upwards and backwards. In the flaccid state this part faces anteriorly. This can cause confusion to the uninitiated!

601 On erotic stimulation, smooth muscle in the fibrous trabeculae and helicine arteries in the three penile corpora relax, owing to parasympathetic activity (pelvic splanchnics S2, 3, 4). The lumina of these arteries enlarge and straighten, allowing blood to flow into the cavernous spaces, engorging and dilating them. The bulbospongiosus and ischiocavernosus muscles compress the venous plexi at the base of the corpora, impeding their flow, further increasing intracorporal pressure, causing the penis to become rigid and erect. Detuminescence, after ejaculation and orgasm, results from the sympathetic constriction of the same smooth muscle (superior hypogastric plexus), and relaxation of the muscles of the superficial perineal pouch.

602 The perineal body is a fibromuscular common tendon situated in the midline 1 cm anterior to the anal canal and behind the vagina (in the female) or bulb of penis (in the male). Eight muscles converge and are attached to this point: the two superficial transverse perinei, bulbospongiosus, the two deep transverse perinei, the anterior fibres of the two levator ani muscles and fibres from the superficial external anal sphincter. In the female it is frequently damaged during delivery; the resulting weakness of the pelvic floor may cause prolapse of the pelvic organs.

603 The bulbourethral glands (Cowper) are found only in the male. They are situated in the deep perineal pouch, one on either side of the urethra. Their ducts, however, pierce the inferior fascia of the urogenital diaphragm (perineal membrane) to enter the superficial pouch and enter the urethra in the bulb of the penis. They are homologous in the female with the greater vestibular glands (Bartholin). They lie one on each side under the posterior parts of the labia majora and bulbs of the vestibule. Their ducts empty into the vestibule immediately distal to the hymen.

604 The superficial inguinal lymph nodes drain the following structures on their respective side:

Lower limb
Gluteal region
Anterior abdominal wall below the umbilicus
External genitalia and perineum
Lower half of the anal canal
Vagina below the hymen
Body of the uterus via lymphatics which run in the broad ligament.

It should be stressed that lymph from the testis does not drain to this group of nodes. The lymphatics from this structure follow its blood vessels to the para-aortic nodes. In a case of enlargement of the superficial inguinal nodes, all the structures above must be inspected to ascertain the site of the primary lesion.

B Applied answers

605 On a lateral view of a male urethrogram, the site of the perineal membrane is at the point where the bulbous urethra narrows abruptly on becoming the narrow membranous portion. The narrowing is caused by the sphincter urethrae and is just proximal to the perineal membrane.

606 An inflamed Bartholin's gland is seen as a red swollen labium majus, especially in the posterior part of the vestibule. The gland itself is the size of a pea and lies under cover of the posterior part of the bulb of the vestibule just lateral to the vaginal wall. The ducts open through this wall at the level of the hymen.

607 The widest part of the male urethra, when distended, is the prostatic portion, which is approximately 3 cm long and fusiform in shape. Surprisingly, the narrowest part is the external urethral meatus, flattened laterally. The lumen has thus spiralled through 90° to create a narrow and well-aimed urinary stream! The membranous urethra (at the site of the perineal membrane) runs a close second.

608 Ejaculation is caused by afferent nervous impulses from the glans penis, causing sympathetic discharge to the smooth muscle fibres of epididymis, ductus deferens, seminal vesicles and prostate gland. Joined by secretions from the bulbourethral glands, the semen is ejected from the penis by

rhythmic contractions of the ischiocavernosus and bulbospongiosus muscles. The sympathetic nerves are from the L1–2 outflow and pass to the pelvic organs via the pelvic plexi and especially the hypogastric plexus or presacral nerve, which lies on the promontory of the sacrum. It lies in the concavity of the sacrum and its close relationship with the posterior rectal wall may cause ejaculatory problems following lower rectal surgery, especially after the operation of abdominoperineal excision of rectum.

609 A ruptured perineal body may have few immediate consequences if not repaired properly. However, women in later life are prone to develop herniation of pelvic contents, leading to prolapse of the uterus, a rectocele or cystocele. This is due to the fact that the perineal body is an important anchor point for many perineal muscles.

610 During a mediolateral episiotomy, the perineal skin, posterior vaginal wall and the bulbospongiosus muscle are cut. It is performed to ease the delivery of a fetus when a perineal laceration seems inevitable. If a tear is allowed to occur spontaneously in any direction, it may damage the perineal body (with effect on pelvic visceral support), the external anal sphincter and the rectal wall (with severe effect on anal continence). Episiotomy thus makes a clean cut away from important structures.

611 For a pudendal nerve block, in the lithotomy position, a skin weal is raised halfway between the anus and ischial tuberosity. A 20 cm needle is passed through the weal and directed towards the ischial spine, already palpated per vaginam. The needle is directed just posterior to the inferior tip of the spine. When complete anaesthesia is required, the genital branches of the genitofemoral and ilioinguinal nerves, and the perineal branches of the perforating cutaneous nerve are also anaesthetised by injecting along the lateral margins of the labia majora.

612 Hypospadias is an abnormal opening in the urethral floor, accompanied by an absent external meatus, found in about 0.2% of male babies. The different degrees of hypospadias include the external meatus opening in the glans, along the ventral

aspect of the penis, or even in the perineum at the base of the penis. It is due to a failure of the urogenital folds, which in development are 'zipped' together from the perineum towards the glans. A failure in this mechanism is an important factor in understanding hypospadias. This condition may be associated with hypogonadism, a cleft scrotum and small penis and thus makes sexing of the child quite difficult.

613 When inserting urinary catheters and sounds, the urethral curves must be considered. Just inferior to the perineal membrane, the spongy urethra is well covered inferolaterally by the erectile tissue of the penile bulb, but a short segment just inferior to the perineal membrane is relatively unprotected posteriorly – its thin, distensible wall here is vulnerable to injury during instrumentation, especially as a near right-angled bend occurs during entry into the deep perineal pouch through the perineal membrane. Hence when the instrument reaches this point, the penis must be pulled down between the thighs. The instrument must never be forced, otherwise a false channel may be created. The gentle anteroinferior downwards curve of the prostatic urethra must also be considered.

614 Attachments of the perineal fascia determine the course of extravasated urine in the superficial perineal pouch, after a hard object such as the crossbar of a push bike, in a 'straddle' injury, suddenly compresses the spongy urethra against the hard subpubic arch. Anterolaterally, the ischiopubic rami attach to the perineal membrane, which roofs the superficial pouch. Posteriorly, the skin and the two layers of perineal fascia fuse into the musculofibrous perineal body, denying access to the anal triangle. Hence, the sheath of the penis fills, overspilling into the loose connective tissue of the scrotum. Urine is unable to pass into the thighs because the superficial perineal fascia (Colles) blends with the deep 'fascia lata' of the thigh, just inferior to the inguinal ligament at the groin skin flexure. However, a pathway exists superiorly, anterior to the pubic bone, where urine can track up a potential space between the deep membranous layer of the superficial fascia of the anterior abdominal wall (Scarpa) and the very thin, but strong layer of deep fascia covering the underlying muscles. This

is because Colles' fascia is continuous with Scarpa's fascia. Theoretically, extravasated urine may reach the axillae and neck!

615 Circumcision is simply removal of the foreskin in the male. Special care must be taken when cutting the frenulum which joins the ventral part of the glans penis to the foreskin, as this contains a small but fairly constant artery. It is interesting to note that circumcision in the male is excellent prophylaxis against cancer of the penis. Female circumcision involving the clitoris and labia is still practised world-wide, and is a cause of obstetric problems.

C Information answers

616 The pelvic diaphragm is formed by the paired levator ani and coccygeus muscles. They meet in a midline raphe, closing the inferior pelvic aperture, except for a gap between the anteromedial edges of levator ani. The pelvic diaphragm separates the pelvic cavity from the perineum. It is slung like a funnel-shaped hammock between the pubic body anteriorly and the coccyx posteriorly, with a lateral origin from the arcus tendineus, or 'white line'. The 'white line' is a thickening of obturator internus fascia running from the posterior aspect of the pubic body to the ischial spine, where the coccygeus continues as the posterior and smaller part of the pelvic floor, originating from the sacrospinous ligament and coccyx. Two parts of levator ani are described. Pubococcygeus (its main part) runs from the pubic body and anterior part of the arcus tendineus to the anococcygeal body and coccyx. A deeper subdivision, the puborectalis, doesn't reach the coccyx, but slings back on itself around the anorectal junction, maintaining the flexure here (90°) which is so important to anorectal continence. Some of its fibres sweep around the prostate (levator prostatae) and the mid-vagina (pubovaginalis), anchoring them to the perineal body. The more posterior second part of levator ani, the misnamed iliococcygeus (it has nothing to do with the ilium in man!) runs from the thinner part of the arcus tendineus and ischial spine to the coccyx and anococcygeal body. Levator ani is supplied by the perineal branches of S3 and S4 (recent studies have shown that the perineal

branch of pudendal nerve does not have a role as previously thought). Coccygeus is supplied by S4 and S5.

617 The pelvic diaphragm is the only striated muscle in the body to acquire a resting tonus during sleep. It has an important role in the support of the pelvic viscera, greatly aided by the fibromusculocutaneous focal point of the perineum, the perineal body, with which it is fused. It thus supports the prostate and indirectly the bladder, and in females, the posterior wall of the vagina and indirectly the uterus. It resists inferior thrusting on the pelvic floor which accompany increases in intra-abdominal pressure, e.g. in lifting, coughing, vomiting, micturition, parturition, defaecation, etc. The anorectal angle supports most of the faecal weight in the rectal ampulla, relieving much of the pressure from the external anal sphincter, with which it is continuous. During parturition, levator ani supports the fetal head while the cervix is dilating to permit delivery.

618 Now that the anal canal is recognised as a separate structure from the rectum, the term 'ischiorectal fossa' has been officially abandoned. The pelvic diaphragm divides the pelvic cavity into two parts, a superior part containing the pelvic viscera (and supporting them), and an inferior wedge-shaped space either side of the anal canal filled with fat, the ischioanal fossae. Their lateral wall is the obturator internus fascia with the pudendal nerve and internal pudendal vessels running anteriorly within it (Alcock's canal) and the ischial tuberosities. Their roof is levator ani, their floor the perianal and perineal skin. Anteriorly, each fossa extends deep to the urogenital diaphragm, paraurethrally, but with no communication with each other. However, posteriorly communication does occur over the anococcygeal body, with the possibility of formation of the horseshoe-shaped bilateral ischioanal abscess.

619 As its name suggests, the external anal sphincter is situated outside the internal anal sphincter. The internal sphincter is a thickening of the circular muscle of the intestinal wall and is composed of smooth muscle. Therefore, its nerve supply is from the autonomic system via the pelvic plexi. The external anal sphincter is composed of striped

muscle and is under voluntary control. It is divided into three parts: subcutaneous, superficial and deep. The subcutaneous part is a muscular ring with no attachments. The superficial part consists of fibres which are attached to the coccyx posteriorly and the perineal body anteriorly, and this helps to anchor the anal canal. The deep sphincter is a ring of muscle which is fused to the puborectalis part of the levator ani, forming the anorectal ring. This structure is vital for continence. Being a voluntary muscle, the external sphincter is supplied by the somatic nervous system, the inferior rectal branch of the pudendal nerve.

620 The midpoint of the anal canal represents the junction between hindgut and body wall. The blood supply of the region reflects this fact. The artery of the hindgut is the inferior mesenteric artery. Its continuation, the superior rectal artery, divides into left and right branches as it reaches the rectum. The right branch divides again into anterior and posterior divisions. These branches supply the rectum and upper part of the anal canal in the position 3, 7 and 11 o'clock. This is with the subject in the lithotomy position. The inferior rectal artery (a body wall artery) is a branch of the pudendal artery, there being one on each side. They supply the lower half of the anal canal. The middle rectal artery – again one on each side – is a branch of the internal iliac artery. They supply only the muscle wall of the rectum. The veins correspond to the arteries, the superior rectal draining to the inferior mesenteric and eventually portal vein, and the inferior rectal to the iliac vein via the internal pudendal. The anal canal, therefore, is a site of portosystemic venous anastomosis.

621 The human rectum, unlike that of other animals, is not straight. The upper and lower parts sit in the midline but the central portion is thrown out to the left. Three 'bends' in the rectum are thus produced; from above downwards, they turn left, right and left. Inside the rectum the 'inside' of each bend is marked by a 'shelf' protruding from the rectal wall into the lumen and produced by all the layers of the wall. These three 'shelves' are the transverse folds or rectal valves (Houston). Their function is unclear, but in sigmoidoscopy they cause a nuisance by obstructing the instrument.

C Applied answers

622 The pubococcygeus, the main part of levator ani, is most likely to be damaged during childbirth. It is obstetrically important because it encircles and supports the urethra, vagina and anus. Injuries to the pelvic fascia and pubococcygeus may result in cystocele (herniation of the bladder into the vagina) or cystourethrocele. A weakened pubovaginalis, stretched and lacerated during parturition, may result in urinary stress incontinence (urinary dribbling on laughing, lifting, etc.)

623 The pectinate line is the demarcation in the anal canal between the embryonic proctodeum (ectoderm) and the embryonic cloaca (endoderm). The hymen is the partition between the vagina and the vestibule. The pectinate line divides the systemic blood supply inferiorly from the portal venous supply superiorly, and in a similar manner, the autonomic nervous plexi supply above this line whilst the somatic inferior haemorrhoidal nerves supply below. The lymphatic drainage of the vagina also has a watershed; below the hymen and the pectinate line the skin of the vagina and anus drain to the superficial inguinal lymph nodes, whilst the upper vagina and anus drain to iliac nodes. Thus both the hymen and the pectinate line are embryonic watersheds with important clinical significance.

624 The exact position of an anal carcinoma will determine the direction of lymphatic spread. Though there is clinically some overlap, a tumour situated below the pectinate line will drain to the groin (superficial inguinal nodes) whilst one above the line will spread to the internal iliac nodes. These latter will of course require pelvic surgery to be removed.

625 An imperforate anus is a failure of the proctodeum and cloaca to meet. The pectinate line probably represents the former site of the anal membrane, which normally breaks down during the seventh week of fetal life allowing the rectum to open into the amniotic sac. This membrane lies at the top of the anal pit, which is lined by ectoderm and formed by a dorsal extension of the urogenital fold. This explains why a shallow anal pit at birth masks the presence of an imperforate anus.

626 During a digital rectal examination in both sexes, the pelvic surfaces of the sacrum and coccyx may be felt, together with the ischial tuberosities and spines. Enlarged internal iliac lymph nodes, pathological thickening of the ureters, swellings of the ischioanal fossae (e.g. abscesses) and abnormal contents of the rectouterine pouch (Douglas) in the female/rectovesical pouch in the male (e.g. fluid) can all be felt. Tenderness of an inflamed vermiform appendix can also be detected if it hangs down in the pelvis. In the male, the prostate and its median sulcus, and spongy urethra (especially if catheterised) may be felt. In the female, the cervix uteri, or a retroverted uterus is easily palpated and may even be mistaken for a rectal neoplasm! The perineal body, and sometimes the ovaries hanging in the pouch of Douglas (after stretching of the broad ligament during pregnancy, with its subsequent laxity) may also be felt. On finger withdrawal, the anorectal ring (junction between external anal sphincter and puborectalis) is felt as a thick band, and more inferiorly the 'submucous space' is felt as a loose section of the anal skin (pecten) just superior to the tightly adherent anal skin over the palpable intersphincteric groove.

627 Haemorrhoids (or piles) are varicosities of the superior haemorrhoidal (rectal) veins. These veins are part of the portal venous system and it is within this region that there is a portosystemic venous anastomosis. Blockage of the portal venous outflow (e.g. due to cirrhosis of the liver) will therefore cause the blood in the portal system to find alternative venous channels. The portosystemic venous anastomoses may manifest as numerous enlarged tortuous veins seen as varices in the oesophagus or haemorrhoids in the rectum. Injection of irritant solutions into haemorrhoids should be performed above the pectinate line and hence into a region innervated by the autonomic nervous plexi, which is painless. Injection below the pectinate line is very painful due to its somatic innervation and you may lose your patient's custom!

628 The lithotomy position is literally that used for 'cutting for stone'. In fact, it is lying supine with the legs abducted and raised in stirrups for easy access to the perineum. It is used mainly for gynaecological and anal procedures. In this rather

undignified position the boundaries of the anal triangle are the coccyx and two ischial tuberosities, with the base of this triangle across the perineum between the tuberosities, the apex being the coccyx.

629 The major content of the ischioanal fossa is dense lobulated fat. This is a common site of both abscess and fistula formation due to the medial border of the fossa being the anal canal. It is therefore possible for fistulae to track into the fossa and form an abscess deep in the fat. External injury in the perianal skin may also lead to deep-seated abscess formation in the fossae. In performing surgery on fistulae and abscesses in this region, it is most important to leave the anorectal ring intact to avoid faecal incontinence.

630 An obturator hernia is the rare occurrence of the abnormal passage of peritoneum with bowel or its mesentery through the obturator canal, where the obturator vessels and nerve pass from the pelvis through the obturator foramen and its closing membrane into the medial thigh to supply the adductor muscles. Pressure on the obturator nerve may cause referred pain to the medial thigh, hip or knee joints.

Cancer of the ovary may invade the peritoneum of the ovarian fossa on the lateral wall of the pelvis, the sensation of which is carried by the obturator nerve, with a similar pattern of pain referral.

D Information answers

631 The ejaculatory ducts pierce the prostate gland and terminate in the prostatic urethra. Their openings are found one on each side on the urethral crest, within the opening of the utricle.

632 The ductus (vas) deferens is a 45 cm long thick-walled smooth muscular tube continuous with the coiled duct of the epididymis at its tail. Ascending in the spermatic cord, it passes through the inguinal canal, hooks around the medial wall of the deep inguinal ring, the inferior epigastric vessels and crosses over the external iliac vessels to enter the true pelvis. Along the lateral pelvic wall it lies external but adherent to the parietal peritoneum, with no other structure intervening between the

two. The ductus crosses the ureter near the posterolateral angle of the bladder, lying firstly superior, then medial to the seminal vesicles and ureter. It enlarges to form its ampulla posterior to the bladder, finally narrowing to join the duct of the seminal vesicles thus forming the common ejaculatory duct.

The tiny artery to the ductus adheres to it, arising from superior or inferior vesical arteries, terminating by anastomosing with the testicular artery, posterior to the testis.

633 The seminal vesicles are a pair of tubes, 10–15 cm long, coiled to form two obliquely placed pear-shaped organs between the fundus of the bladder and rectum, superior to the prostate. They do not store sperm (as their name implies), but secrete a thick alkaline fluid, rich in fructose, that mixes with the sperm as it passes into the common ejaculatory ducts and urethra. These secretions, expelled when the seminal vesicles contract under sympathetic stimulation during orgasm, provide most of the seminal volume.

634 The innervation of the urinary bladder and its connecting urogenital structures is by the autonomic and somatic nervous systems. Parasympathetic fibres via the pelvic splanchnics (S2, 3, 4) are motor to the detrusor muscle of the bladder and control the internal sphincter at the bladder neck, with which it is continuous. Sympathetic fibres, relayed through the superior hypogastric plexus (T11, T12, L1, L2), are motor to a continuous sheet of smooth muscle comprising the ureters, trigonal and urethral crest muscle and the smooth muscle of the prostate, seminal vesicles, ductus deferens and the duct of the epididymis. The pudendal nerve (somatic S2, 3, 4), via its perineal branch, is motor to sphincter urethrae and sensory to the spongy urethra and glans penis.

635 The internal iliac artery is important because it supplies the viscera and walls of the pelvis. It arises at the bifurcation of the common iliac artery in front of the sacroiliac joint. It descends to the greater sciatic foramen, where it divides into anterior and posterior divisions. The anterior division gives rise mainly to visceral branches. These are the umbilical artery (which, when obliterated, forms the medial umbilical ligament), the superior

and inferior vesical arteries to the bladder, the uterine artery, the middle rectal artery to the muscle of the lower rectum and the obturator artery. The anterior division ends in the internal pudendal and inferior gluteal arteries. The posterior division gives rise to parietal branches. These are the iliolumbar, lateral sacral and superior gluteal arteries.

636 The normal prostate is described as having two capsules, and the pathological gland three. Both statements are incorrect. The prostate has only one true capsule, which is a thin fibrous envelope around the gland. The false capsule, outside the true capsule, is separated from it by the prostatic venous plexus. It consists of thickened pelvic fascia. Pathologically, an adenoma of the prostate will compress the remainder of the gland against the true capsule. This rim of squashed tissue inside the true capsule is the pathological capsule of the prostate.

637 For the purposes of description, the rectum may be divided into thirds: upper, middle and lower. The relationship of the peritoneum is different in each third. In the upper third the rectum is covered by peritoneum anteriorly and on both sides, and only posteriorly is it absent. In the middle third, the peritoneum covers only the anterior aspect of the rectum, leaving both sides and the posterior aspect deficient. In the lower third, the rectum has no relationship with the peritoneum, as the latter has been reflected onto the anterior pelvic organs.

638 The prostatic (first) part of the urethra, 3 cm in length, has a gentle anterior concave curve, and its middle section is the widest and most distensible part of the urethra, despite being inside the solid prostate (although it is contracted except on the passage of fluid). The posterior wall has several notable features, the most prominent being the median longitudinal urethral crest (L. *verumontanum* = mountain of truth), with a groove on either side, and the prostatic sinuses, where most of the prostatic ductules open. On the middle of the crest is a rounded eminence, the seminal colliculus (L. = little mound), from which a slit-like orifice leads into a small vestigial cul-de-sac, about 5 mm in length, the prostatic utricle (L. small leather bag), a remnant of the uterovaginal canal in the male

embryo (hence the old names of 'vagina masculina' and 'uterus masculinis'). In an uncommon type of intersex it bleeds cyclically, a very rare cause of haematuria! On each side of the orifice of the prostatic utricle are the minute openings of the two common ejaculatory ducts – they are very difficult to see, but can be cannulated.

639 The trigone is a smooth triangular area situated on the internal aspect of the posterior wall (base) of the bladder. The three angles of the trigone are the two ureteric orifices superolaterally and the internal urethral orifice inferiorly. Between the two ureteric orifices is a ridge of muscle – the interureteric bar. The trigone, unlike the rest of the organ, does not distend as it fills. Pain receptors are numerous in this region.

640 The prostatic urethra is not circular in cross-section but kidney-shaped, consisting of a posterior midline ridge with a gutter on either side. The gutters constitute the prostatic sinus; the ridge is the verumontanum or urethral crest.

D Applied answers

641 The female genital duct is called the paramesonephric duct (Müller). During male development most of this duct regresses, leaving only the appendix testes and prostatic utricle in the male adult. Both of these may develop pathologically, the utricle producing a cystic swelling in the region of the verumontanum, whilst the appendix may undergo torsion and be misdiagnosed as acute epididymitis or even appendicitis due to the location of referred pain.

642 The prostatic venous plexus receives the deep dorsal vein of the penis and drains into the vesicular and hypogastric veins; eventually these drain into the internal iliac veins. The plexus is both valveless and thin-walled, and it has been shown that there are connections with the vertebral venous plexus (Batson). It is suggested that the frequent spread of prostatic malignancy to the vertebral column is through this connection.

643 The urethra passes through the prostate gland itself, this portion being the widest 3 cm of the male

urethra. Enlargement of the gland, and in particular its median lobe, causes retention of urine due to compression of the urethra, as well as an urge to frequent micturition due to pressure on the bladder neck.

644 A normal vasogram will reveal the ductus deferens (vas deferens) coming from the scrotum and crossing the pelvic cavity. It will often also show the ampulla of the vas lying just medial to the seminal vesicle and the two ejaculatory ducts opening into the prostatic portion of the urethra at the level of the verumontanum.

645 Pain located at the tip of the penis is often referred pain from the trigone of the bladder. As this pain is associated with the end of micturition and haematuria, it would be reasonable to suggest that it is due to a bladder stone irritating the trigone as the bladder empties.

646 As the bladder fills, it rises over the symphysis pubis like the sun over the horizon (up and to the level of the umbilicus after a heavy drinking session!), expanding into the extraperitoneal fat lifting several centimetres of parietal peritoneum from the anterior abdominal wall. Now lying adjacent to the wall without intervening peritoneum, surgical approach or suprapubic catheterisation is extraperitoneal. The bladder so filled is prone to rupture by a lower abdominal blow, or by a fracture of the superior pubic rami from a fall on the feet, or a car crash crushing the pelvis in an anteroposterior plane. In such cases, its superior peritoneal covering is frequently torn with intraperitoneal extravasation of urine. A posterior rupture causes passage of urine extraperitoneally into the perineum.

647 The size, shape and position of the bladder vary with the volume of urine it contains, but in the normal adult one would expect to palpate or to percuss it only if it is well distended. In babies and young children, however, it tends to be a relatively superficial organ situated largely in the abdominal cavity.

648 This is strictly the new name for a vasectomy (Nomina anatomica, 1989), the common method of sterilising males which involves two small inci-

sions made each side in the superior part of the scrotum. Usually each duct is exposed and tied in two places just before the spermatic cord enters the superficial inguinal ring, and the portion between the sutures is excised. Therefore, although spermatogenesis continues, ejaculatory fluid from the seminal vesicles, prostate and bulbourethral glands contain no sperm. To reach the ductus, the three major coverings of the cord are incised, namely the external spermatic fascia derived from the external oblique aponeurosis, the cremasteric fascia and muscle, derived from the internal oblique muscle and the internal spermatic fascia derived from the transversalis fascia.

649 The position of the prostate gland depends on the fullness of the bladder – a full bladder displaces the gland inferiorly, making it easier to palpate. A normal prostate is an elastic, symmetrical swelling with a midline sulcus: a malignant prostate feels hard, nodular and may loose this sulcus. The sulcus may also often be lost in benign hypertrophy.

650 Two planes of endopelvic fascial condensations greatly aid the surgeon in abdominoperineal resection of the rectum. The rectovesical septum (Denonvilliers) enables easier separation of the prostate and urethra from the rectum, thus reducing the risk of damage to these structures. Waldeyer's fascia loosely attaches the rectum and sacrum to each other, but forms a relatively impermeable barrier to the direct spread of rectal cancer posteriorly. The extensive anastomosis of arteries in all three layers of the rectal wall enables the clamping of the superior rectal artery to not affect rectal viability in surgery of localised lesions.

651 Ejaculation is a reflex action during the climax of passage of erotic thoughts or afferent impulses from the glans penis. It consists of two phases: the emission of sperm into the urethra following sympathetic discharge to the smooth muscle of the ductus deferens, seminal vesicles and prostate; and ejaculation proper with clonic spasm of bulbospongiosus and ischiocavernosus muscles (perineal branch of pudendal nerve). After anorectal surgery, damage to the pelvic splanchnic nerves (which give off branches passing forwards around the rectum) cause failure of erection, while injury to the sym-

pathetics (from L1, 2 via the hypogastric plexus and its branches anterior to the sacrum) can cause ejaculatory problems, especially after abdominoperineal excision of the rectum.

E Information answers

652 The ovary is pulled into its adult position in the pelvis by the gubernaculum ovarii. The ovary originates from the upper lumbar region and, as it descends, it brings its blood, nerve and lymphatic supplies with it. Thus, in the adult, the ovarian arteries arise from the abdominal aorta immediately below the origin of the renal arteries (L1–2). The ovarian veins normally drain on the right into the inferior vena cava and on the left into the left renal vein (at the level L1–2). Furthermore, the ovarian nerves are derived from the renal and aortic plexi and accompany the blood vessels into the pelvis.

653 The round ligament of the uterus and the ligament of the ovary are the remains of the gubernaculum. In the male the gubernaculum pulls the testis from the lumbar region into the scrotum. In the female the descent of the ovary is arrested in the pelvis, but the gubernaculum may still be seen running from the pelvis, through the inguinal canal and into the labium majus (the homologue of the scrotum).

654 The homologous structures to the prostate in the female are the paraurethral glands of Skene, a greater number of these being alongside the superior half of the female urethra, an area corresponding to the prostatic part in the male (there is no spongy equivalent). These glands have a common paraurethral duct which opens, one on each side, near the external urethral orifice.

655 The uterus is normally described as anteverted and anteflexed. The long axis of the uterus is normally at an angle of some 90° to the long axis of the vagina, the uterine cervix thus entering the anterior vaginal wall. This is termed anteversion. Furthermore, the body and fundus of the uterus form an angle of some 160° with the cervix, so that the uterus sits snugly on the bladder (anteflexion). In a small percentage of normal women, as well as pathologi-

cally, these relationships may be lost. Retroversion involves the uterus pointing towards the rectum with the vagino-uterine angle exceeding 180°. Reversal of the angle between cervix and uterine body to greater than 180° would be termed retroflexion.

656 The uterine tube is some 10 cm long. It is found in the upper free border of the broad ligament and connects, via its lumen, the cavity of the uterus with the peritoneal cavity. It consists of four parts; these are, from lateral to medial – infundibulum, ampulla, isthmus and interstitial part. The infundibulum is the funnel-shaped lateral extremity; it terminates in a number of finger-like processes (fimbriae), one or more of which clasp the ovary. The ampulla is the dilated portion of the tube adjacent to the infundibulum, where normal fertilisation occurs. The isthmus is the narrowest part of the tube and is found just lateral to the uterus. The interstitial part of the tube is that part passing through the uterine wall and joins the isthmus to the cavity of the uterus.

657 The drainage of lymph from the uterus is different in the various regions of the organ, viz. fundus, body and cervix. The fundus sends its lymph, together with that from the uterine (Fallopian) tube and ovary, along the ovarian vessels to the para-aortic nodes. The body drains laterally via the broad ligament to the lateral wall of the pelvis and nodes lying next to the external iliac vessels, the external iliac nodes. Part of the lymph from the body drains along lymphatics in the round ligament of the uterus to the superficial inguinal nodes. The cervix drains laterally via the broad ligament to the external iliac nodes, posterolaterally with the uterine vessels and posteriorly via lymphatics in the rectouterine fold round either side of the rectum to the sacral nodes.

658 The urachus is the allantoic diverticulum running from the cloaca into the umbilical cord. In the adult it is represented as a fibrous strand running from the apex of the bladder to the umbilicus – the median umbilical ligament. In the fetus the umbilical artery runs, one on each side, from the anterior division of the internal iliac artery to the umbilicus. In the adult they are obliterated and the resulting fibrous strand is the medial umbilical ligament.

659 The broad 'ligament' is a peritoneal fold extending from the sides of the uterus to the lateral pelvic walls and floor (just as if a sheet were draped over the uterus and its tubes, which lie in its upper free border). The ovary is attached to its posterior aspect by the mesovarium, with the ovarian 'services' running through it into the broad ligament, then into its superior extension, the 'suspensory ligament of the ovary' (another fold of peritoneum draping over them) to reach the lateral pelvic wall. Within the broad ligament course the ovarian and the round ligament of the uterus, both derived from the embryonic gubernaculum: they are continuous within the lateral walls of the uterus. The broad ligament also contains extraperitoneal connective tissue and smooth muscle (parametrium). The mesosalpinx is that part of the broad ligament between the ovary, its ligament and the uterine tube: it sometimes contains remnants of the cranial and caudal ends of the mesonephric tubules of the temporary embryonic kidney (mesonephros), called the 'epoöphoron' and 'paroöphoron' respectively. Remnants of the cranial end of the mesonephric (Wolffian) duct, forming the ductus epididymis in males, can remain as vesicular appendages on the infundibulum of the uterine tube: more inferiorly, that part forming the deferential and ejaculatory ducts can persist in females as the 'duct of Gartner', lying between the layers of the broad ligament, alongside the lateral walls of the uterus (or vagina). These vestigial structures occasionally swell with fluid to form cysts.

660 The cervix uteri enters the anterior wall of the vagina and extends a short distance into it. Thus the upper part of the vaginal lumen surrounds the cervix like a gutter. This is the fornix. Although a continuous structure, the fornix is descriptively divided into anterior, posterior and right and left lateral fornices for purposes of orientation. The posterior fornix is deeper than the anterior owing to the fact that the cervix enters the anterior vaginal wall due to the usual anteverted position of the uterus.

661 In the male, the peritoneum passes from the anterior abdominal wall onto the superior surface of the bladder. It runs down the posterior surface of the bladder for a short distance and is then reflected

onto the middle third of the rectum, forming the rectovesical pouch as it does so. It then runs superiorly to form the parietal peritoneum of the posterior abdominal wall.

In the female the uterus is interposed between bladder and rectum. The peritoneum therefore runs from bladder to uterus, forming the uterovesical pouch, and thence from uterus to rectum, forming the rectouterine pouch (Douglas).

662 The uterus is supported by the levator ani muscle and various condensations of pelvic fascia containing muscle termed ligaments. The levator ani forms the bulk of the pelvic diaphragm and supports the pelvic viscera above it. It supports the uterus indirectly by supporting the vagina. Condensations of pelvic fascia, namely the cardinal, pubocervical and uterosacral ligaments are situated above levator ani and are attached to the cervix and upper vagina. The cardinal ligament runs to the lateral pelvic wall on each side. The pubocervical ligaments run from the pubic bone, and skirt either side of the bladder and bladder neck to reach the cervix uteri. The uterosacral ligaments form two bands on the ipsilateral aspect of the pouch of Douglas as they pass lateral to the rectum and attach to the lower sacrum.

E Applied answers

663 The close relationships between the ureter and two arteries in the pelvis make its accurate identification essential in gynaecological surgery, and can be the cause of disturbed sleep for the surgeon in uncertain cases, retrospectively! The uterine artery crosses anterior and superior to the ureter near the lateral aspect of the posterior vaginal fornix, where its pulsation can be felt. Here, the ureter is in danger of being inadvertently clamped or severed during hysterectomy (especially on the left), when the uterine artery is tied off: its point of crossing is 2 cm superior to the ischial spine. During ovariectomy, the ureter is also vulnerable to injury when the ovarian vessels are being tied off – both lie very close together as they cross the pelvic brim.

664 Obstruction of the ureters by calculi occurs most often where they cross the external iliac artery and

pelvic brim, and also where they pass obliquely through the wall of the bladder. The presence of calculi can be confirmed in the majority of cases by plain abdominal X-ray (80%) or intravenous urogram, and can be removed by open surgery, cystoscopy or lithotripsy.

665 During a vaginal examination it is theoretically possible to palpate a ureteric stone through the lateral fornices. It is here that the ureters lie in the base of the broad ligament just lateral to the vaginal and uterine arteries which anastomose along the upper vaginal walls.

666 Usually preceded by a speculum examination, a bimanual pelvic examination can feel the external os of the cervix, the fornices and can be used to assess the size, position and shape of the uterus. A normal ovary may often be felt but the uterine tubes are palpable only if enlarged or thickened by disease. Examination of the posterior fornix may also reveal any abnormalities in the pouch of Douglas and often a normal ovary. The size of the bony pelvis may also be assessed, and a narrowed diagonal conjugate diagnosed if the examining fingers can reach the sacral promontory.

667 Free 'overspill' in the pouch of Douglas is an important sign, indicating patent uterine tubes, as distinct from loculated spill as a result of damage caused by pelvic inflammatory disease. The hysterosalpingogram is a commonly used test in the investigation of infertility.

668 If an embryo is destined to become female, the mesonephric tubules degenerate into the vestigial paroöphoron and epoöphoron of the broad ligament. The mesonephric duct (Wolffian) remnants are the epithelial cysts from the broad ligament to the vestibule and are called the canals of Gartner. These may cause cystic swellings requiring surgery.

669 The amateur abortionist often does not know the close relationship of the posterior fornix with the pouch of Douglas and the anteverted position of the uterus. In most women with an anteverted uterus, the posterior vaginal wall is longer than the anterior wall and in a direct line with the posterior fornix. Hence the aborting instrument slides along

the posterior wall, through the thin wall of the posterior fornix and directly into the peritoneal cavity, with resultant peritonitis.

670 At the time of ovulation some women experience intermenstrual pain (mittelschmerz) due to the stretching of the ovarian wall. This is experienced as referred pain to the dermatome of the innervation of the ovary. The ovary is supplied from T10 fibres and so the pain is paraumbilical (cf. testis).

671 To explain an abdominal pregnancy one must realise that the ovum is expelled into the peritoneal cavity, only later to be diverted into the uterine tube by the fimbriae. If, as very rarely happens, the ovum is fertilised inside the peritoneal cavity by a sperm which has swum out of the peritoneal opening near the fimbria, then implantation can occur inside the peritoneal cavity. The other, more likely, occurrence is that during an abortion or rupture of a tubal ectopic pregnancy, the trophoblast reimplants itself in the broad ligament or pelvic peritoneal cavity. These abdominal pregnancies may proceed to near term before they are discovered.

672 The uterus is supported or suspended by three groups of ligaments: the pubocervical anteriorly, the cardinal (Mackenrodt) laterally and the uterosacral posteriorly. These ligaments also help maintain the uterus in its correct position. Their stretching may lead not only to retroversion but, if they are not strong enough to counterbalance increases in intra-abdominal pressure, also to a gradual descent of the uterus. This can result in a complete prolapsed uterus (proccidentia).

673 The lymphatic drainage of the ovary is to the para-aortic nodes on the posterior abdominal wall at the level of the renal vessels.

LOWER LIMB

A Information answers

674 There are three bursae associated with the gluteus maximus which separate this large muscle from underlying structures: the trochanteric bursa, separating it from the lateral side of the greater

trochanter; the ischial bursa, superficial to the ischial tuberosity (weaver's or jeep bottom is ischial bursitis); and the gluteofemoral bursa which separates the gluteus maximus from the proximal attachment of vastus lateralis.

675 The fibrous capsule of the hip joint attaches proximally around the edge of the acetabular labrum and transverse acetabular ligament. Distally it attaches to the anterior intertrochanteric line and greater trochanteric root of the femur and posteriorly to the femoral neck halfway up from the intertrochanteric crest – this allows the obturator externus tendon to glide over its bursa (a herniation of the synovial membrane lining the capsule of the hip joint underneath this posterior attachment). Most capsular fibres spiral to the lateral part of the intertrochanteric line; some deep fibres form the zona orbicularis around the femoral neck, acting like a constrictive collar that helps to hold the femoral head in the acetabulum. Some deep longitudinal fibres form retinacula which reflect superiorly along the femoral neck as long bands that blend with its periosteum. They cover ascending blood vessels from the trochanteric anastomosis that supply the head and neck of the femur.

676 Three intrinsic ligaments strengthen the capsule of the hip joint, each from one of the three components of the hip bone.

The iliofemoral ligament (Bigelow) is the strongest, shaped like an inverted Y, running from the anterior inferior iliac spine and acetabular margin, to the intertrochanteric line of the femur. This helps to prevent joint overextension during standing while 'screwing' the femoral head into the acetabulum. The ischiofemoral ligament, from the ischial portion of the acetabular rim, spirals superolaterally to insert mostly into the joint capsule, with the minority of its fibres continuing to insert medial to the base of the greater trochanter (therefore 'ischiocapsular' ligament would be a better name). The weaker pubofemoral ligament, from the pubic part of the acetabular rim and iliopubic eminence, blends with the medial part of the iliofemoral ligament, preventing overabduction of the joint.

The ligamentum teres (ligament of the head of the femur), is a weak 3.5 cm intracapsular band from the margins of the acetabular notch and transverse acetabular ligament to the fovea of the femoral head. It is stretched on adduction and lateral rotation of the hip joint. It has an important role in stabilising the infant's joint before walking, and its accompanying artery (a branch of the obturator artery) is a significant source of perioperative bleeding in hip replacement procedures.

677 The three 'guy ropes' stabilising the hip bone from its furthest separated points give attachment to the upper subcutaneous surface of the tibial shaft. Here the three tendons of sartorius (ilium, femoral nerve), gracilis (pubis, obturator nerve) and semitendinosus (ischium, sciatic nerve) insert in that order, from before backwards. At the tibial insertion, they are separated by a bursa, the 'bursa anserina' (L. goose's foot), deep to the flattened sartorius tendon.

678 The greater and lesser sciatic notches are converted into foramina by the sacrotuberous and sacrospinous ligaments. The sacrotuberous ligament is very strong. It is attached to the posterior inferior iliac spine and the lateral part of the sacrum and coccyx. It runs to the ischial tuberosity. It is supposed to be the phylogenetically degenerated tendon of the long head of biceps femoris. The sacrospinous ligament lies on the pelvic aspect of the sacrotuberous ligament and is triangular in shape. Its base is attached to the lateral aspect of the sacrum and coccyx, and its apex to the ischial spine. Phylogenetically it is the degenerated posterior aspect of coccygeus. These ligaments are important for the stability of the pelvis, for they prevent the backward rotation of the sacrum when the upright posture is assumed.

679 The structures which exit from the pelvis and enter the gluteal region through the greater sciatic foramen below the piriformis muscle are: the sciatic nerve, the nerve to quadratus femoris, the inferior gluteal nerve and vessels, the posterior cutaneous nerve of the thigh, the nerve to obturator internus, the pudendal nerve and the internal pudendal vessels.

The sciatic nerve curves laterally and downwards to pass midway between the ischial tuberosity and greater trochanter into the thigh. The sciatic nerve hides the nerve to quadratus femoris, a branch of the sacral plexus, which also supplies the inferior gemellus. The inferior gluteal nerve (L5, S1, 2) arises from the dorsal divisions of the sacral plexus and supplies gluteus maximus, accompanied by the inferior gluteal vessels. The posterior cutaneous nerve of the thigh supplies skin in the gluteal region and back of the thigh and leg. The nerve to obturator internus crosses the ischial spine and enters the ischioanal fossa via the lesser sciatic foramen. It supplies obturator internus and the superior gemellus. The pudendal nerve and internal pudendal vessels also cross the ischial spine, medial to the nerve to obturator internus, to enter the ischioanal fossa for distribution.

680 Three structures leave the pelvis via the greater sciatic foramen, cross the ischial spine and enter the ischioanal fossa by entering the lesser sciatic foramen. They are, from medial to lateral: the pudendal nerve, the internal pudendal vessels and the nerve to obturator internus.

681 The sciatic nerve is derived from roots L4–S3 of the sacral plexus. It leaves the pelvis via the greater sciatic foramen below the piriformis muscle, to lie in the gluteal region. It curves downwards and laterally into the thigh, passing halfway between the ischial tuberosity and greater trochanter, lying successively on the superior gemellus, obturator internus, inferior gemellus and quadratus femoris. In the thigh it lies on adductor magnus, passing deep to the long head of biceps femoris. In the lower third of the thigh, it divides into the tibial nerve, which passes medially, and the common peroneal nerve, which passes laterally. The tibial nerve is derived from ventral divisions of the sciatic plexus, the common peroneal from dorsal divisions.

682 The ligaments of the hip joint are the three capsular ligaments joining the hip bone with the femur (iliofemoral ligament, ischiofemoral ligament and pubofemoral ligament) and two others – the transverse acetabular ligament and the ligamentum teres.

The strong iliofemoral ligament (Bigelow) is the shape of an inverted Y. Its base is attached to the anterior inferior iliac spine and its two limbs to the intertrochanteric line on the anterior of the femur. The ischiofemoral ligament forms a spiral between the ischium and the greater trochanter. The pubofemoral ligament is triangular. Its long base is attached to the superior pubic ramus, the apex to the inferior part of the intertrochanteric line.

The transverse acetabular ligament is formed by the labrum acetebulare as it spans the acetabular notch, converting it into a tunnel. The ligamentum teres is covered in synovial membrane. It runs between the pit of the head of the femur and the acetabular notch. It transports a small blood vessel to the head of the femur.

Extension of the hip to 15° results in tightening of the iliofemoral ligament, which therefore limits hip extension to this degree.

683 The principal abductors of the hip are gluteus medius and gluteus minimus. Tensor fascia latae and sartorius also abduct the hip, among other actions.

If the hip bone is fixed, the gluteus medius and minimus will abduct the whole limb. If, however, the femur is fixed, the pelvis will be tilted towards the side of the contracting muscles, thereby reversing the role of origin and insertion. This movement is vital in walking. If the right leg is taken off the ground, the pelvis tends to fall to the unsupported right side (Trendelenburg gait). The left abductors contract, however, tilting the pelvis to the left and counteracting this drop. In fact, they overcompensate so that the pelvis is tipped a little to the left. This allows the right foot to clear the ground as the limb is swung forward in taking a pace.

684 The small lateral rotator muscles of the hip are: piriformis, obturator internus and its two gemelli, quadratus femoris and obturator externus.

Piriformis (L. pear-shaped) arises from the anterior surface of the second, third and fourth sacral vertebrae. It passes through the greater sciatic notch to its insertion on the greater trochanter. It forms a convenient reference point for other structures in the gluteal region. Obturator internus arises from the pelvic aspect of the obturator membrane and the surrounding bone. Its tendon passes through the lesser sciatic foramen, a bursa intervening as it winds round the ischium, and onto the greater trochanter. The gemelli, superior and inferior, run above and below the obturator internus tendon to the greater trochanter; the superior arises from the spine of the ischium, the inferior from the ischial tuberosity. The square quadratus femoris arises from the ischial tuberosity and inserts on the quadrate tubercle on the intertrochanteric crest of the femur. With the obturator internus and gemelli above it, it forms a 'bed' for the sciatic nerve. Obturator externus arises from the outside of the obturator membrane and the surrounding ischium and pubis. Its tendon is inserted into a pit on the medial aspect of the greater trochanter.

A Applied answers

685 The gluteal gait (Trendelenburg) is the waddling motion seen when the gluteus medius and minimus and tensor fascia latae (the 'deltoid of the hip joint' separated into its components) fail in their supportive and steadying effect on the pelvis while walking. Normally, the side of the pelvis opposite to that of the rising foot is held in abduction by these muscles so that the pelvis tilts above its normal plane to enable that foot to clear the ground during the swinging phase of that leg. Poliomyelitis, postoperative superior gluteal nerve injury and congenital posterior hip dislocation all affect the action of these muscles.

686 Traumatic hip dislocation is uncommon because the joint is so strong and stable. In a car crash, with the knee striking the dashboard with a flexed, adducted and medially rotated hip joint, the femoral head has, as its posterior relation, more capsule and less bone. This results in a posteroinferior capsular rupture and passage of the femoral head through this tear, over the posterior acetabular margin to lie on the gluteal surface of the ilium. Often there is an accompanying fracture of the margin (fracture dislocation), the head taking with it the acetabular fragment and labrum. The sciatic nerve (L4–S3) is a posterior relation of the hip joint and is therefore at risk.

687 The patient lies prone, or on his side, with the buttock muscles relaxed. The safe area for injection, if the buttock is to be used at all, is situated in the upper outer quadrant of the true anatomical

buttock, and not just of the 'cosmetic' buttock! This is best located by marking a line from the greater trochanter of the femur to the dimple which marks the posterior superior iliac spine. Above this line is a safe area, and the lateral part of this region is often used for intramuscular injections. Nowadays, due to the dangers of hitting the sciatic or even the gluteal nerves, the tensor fasciae latae or vastus lateralis muscles are often the sites of choice, particularly in an uncooperative patient.

688 If the L4/5 disc has prolapsed, the nerve most likely to be affected is the L5 nerve root. Similarly, L5/S1 disc protrusion is most likely to cause problems with the S1 nerve root. Remember, that in this region, the spinal nerves exit from the intervertebral foramina below their numbered vertebral body, but usually above the level of the intervertebral disc, so that the more common lateral protrusions affect the spinal nerve below, whereas larger midline posterior disc protrusions may affect three or more of the lower roots. The pain will be worst in the same dermatomes, and these may show loss of sensation.

The L5 dermatome involves the lateral part of the lower leg, dorsum of the foot and medial three toes, as well as the heel and plantar surface of the middle toes. The S1 dermatome extends from the buttock down the posterior surface of thigh and calf, and extends laterally to end in the region of the lateral malleolus and little toe on both dorsal and plantar surfaces. Motor function may also be affected, especially dorsiflexion (L4, 5) plantar flexion (S1, 2) or eversion (S1). In L5/S1 disc protrusion, examination of the calf muscles may reveal wasting, and the ankle jerk may be diminished or even absent.

689 The blood supply of the head of the femur is from three routes. First, from the circumflex femoral arteries ascending in the synovial retinaculum along the femoral neck; second, from nutrient arteries ascending from the marrow of the shaft; and lastly, a small contribution along the ligament of the head (ligamentum teres) which is often a branch from the obturator artery. Because most of these arteries run upwards along the neck, and because the capsule and retinacular arteries join the bone low down the neck, a fracture through the neck itself is likely to interrupt the blood supply. Subcapital fractures, particularly, are prone to avas-

cular necrosis, as the artery of the ligament of the head is not sufficient to keep the bone alive. A fracture through the trochanters, however, barely has an effect on the blood supply to the head, as the circumflex femoral vessels enter proximal to the break.

690 Shortening of the fractured femur is due to the strength of the longitudinally lying muscles, especially the quadriceps and hamstrings, as well as the adductor group which pull the limb superomedially. Also, instead of rotating about an axis between hip and knee, the femur now rotates about the axis of its shaft. This means that the upper adductors and iliopsoas now rotate the femur laterally, instead of medially as before.

691 Pure extension of the hip is normally limited to some 15–30° by the iliofemoral ligament and iliopsoas tendon. The movement, easily mistaken for extension, is, in fact, a combination of extension, pelvic tilt, abduction and lateral rotation. In testing extension, therefore, it is important, first, to fix the pelvis and, second, to perform the movement with the knee flexed. The hamstrings and gluteus maximus produce extension.

B Information answers

692 Starting superiorly the cutaneous innervation of the thigh consists of the subcostal nerve (T12) anterior to the greater trochanter, the femoral branch of the genitofemoral nerve (L1) inferior to the middle third of the inguinal ligament, and the iliohypogastric nerve (L1) over the mons pubis. The lateral femoral cutaneous nerve (L2, 3) – a direct lumbar plexus branch – supplies the thigh laterally. Entrapment of this nerve under the inguinal ligament or irritation by the pull of the overlying transversalis fascia on the posterior abdominal wall presents as the syndrome of meralgia paraesthetica, with tingling and pain along its cutaneous distribution. Direct branches of the anterior division of the femoral nerve – the intermediate and medial femoral cutaneous nerves – supply most of the anterior thigh and prepatella skin. The anterior division of the obturator nerve (L2, 3, 4) gives sensation to the medial thigh. Posteriorly, from the lower third of the buttock

down the back of the thigh and over the popliteal fossa to between the heads of gastrocnemius, small branches of the posterior femoral cutaneous nerve (S2, 3) pierce the fascia lata to give cutaneous innervation. This nerve is the longest sensory nerve in the body, and the only one to run under deep fascia!

693 This thickened layer of deep fascia invests the thigh muscles like ladies' tights, preventing them from bulging excessively and inefficiently on contraction, and also helping to maintain intramuscular pressure for venous return. Superior attachments to the limb root include the inguinal ligament, external lip of iliac crest, posterior surface of sacrum and coccyx, sacrotuberous ligament, ischial tuberosity, margin of pubic arch, pubic body and tubercle. Laterally, from the tubercle of the iliac crest to the tibia, it is thickened as the strong iliotibial tract. This receives tendinous reinforcements from three-quarters of gluteus maximus posteriorly, counteracted by the anterior pull of the tensor fascia latae muscle. The postural function of the fascia lata is seen to permit the 'action at a distance' of gluteus maximus on the already extended knee joint, enabling the vasti muscles to relax, when it acts anterior to the centre of the knee joints, keeping them extended.

694 The saphenous opening is a deficiency in the fascia lata just inferior to the inguinal ligament. It is 4 cm long and 2 cm wide, with its centre 3 cm inferolateral to the pubic tubercle. The smooth medial margin is contrasted by a sharp crescentic edge, the falciform margin, laterally: the two edges are joined by the fibrofatty cribriform fascia which is pierced by the long saphenous vein and efferents from the superficial inguinal nodes. A saphena varix may form here – this is not to be confused with a femoral hernia!

695 The femoral artery enters the femoral triangle deep to the midpoint of the inguinal ligament, lateral to the femoral vein, where it can be compressed against the femoral head to arrest haemorrhage in lower limb trauma. Descending on iliopsoas, pectineus and adductor longus, it bisects the floor of the triangle and at its apex runs deep to sartorius in the adductor canal. In the canal, the vein is now posterior to the artery. Both

run posteromedially through the adductor hiatus into the popliteal fossa. The surface marking of the femoral artery, with the hip slightly flexed and laterally rotated, is from the midpoint between pubic tubercle and anterior superior iliac spine, along the superior two-thirds of a line running towards the adductor tubercle. The tubercle is often palpated on the lower medial aspect of the femoral shaft.

696 The cruciate anastomosis is the union of the medial and lateral circumflex femoral arteries with the inferior gluteal artery above, and the first perforating branch of profunda femoris below, posterior to the upper femoral shaft. It provides an alternative blood route for the distal lower limb should the femoral artery be blocked or surgically ligated.

697 Enlargement of the superficial inguinal lymph nodes occurs from diseases in the areas they drain. Minor sepsis and abrasions of the lower limb give slight enlargement in otherwise healthy people. Malignancies and infection of the external genitalia, perineal abscesses, malignancy of the lower anal canal and of the uterus or cervix (spreading via the lymphatics surrounding the round ligament), and breast metastases via superficial lymphatics of anterior abdominal wall (by abnormal retrograde flow from obstructed normal axillary and thoracic routes) can all give significant enlargement. Note that the testes drain to the para-aortic nodes of the posterior abdominal wall and not to these nodes.

698 Quadriceps femoris has four heads, i.e. vastus medialis, vastus intermedius, vastus lateralis and rectus femoris. Three arise from most of the femoral shaft.

Vasti lateralis and medialis arise from the lateral and medial lips of the linea aspera on the posterior aspect of the shaft, and intermedius from the shaft anterior to these lips. The fourth head, rectus femoris, originates from the anterior inferior spine of the ilium (straight head) and from a slip just superior to the acetabular margin (reflected head). The rectus, anterior to the vasti, thus flexes the hip joint as well as extending the knee joint via their common quadriceps tendon, patella, and the patella ligament attached to the tibial tuberosity. If quadriceps is paralysed, the erect posture can be maintained as the body's centre of gravity lies anterior to the knee joint, tending to overextend the knee – patients in this situation tend to walk pressing on the distal end of their thigh to prevent knee flexion.

699 The femoral triangle is situated on the upper medial aspect of the front of the thigh. Its boundaries are the inguinal ligament superiorly, the sartorius laterally and the medial border of the adductor longus muscle medially. Its floor is composed, of adductor longus, pectineus and iliopsoas (from medial to lateral). Its roof is the fascial layers and the skin of the thigh. It contains laterally the femoral nerve, splitting into its terminal branches, and medially a funnel-shaped fascial structure, the femoral sheath. This structure contains from lateral to medial, the femoral artery, femoral vein and femoral canal. The canal is full of fat and lymphatics. The lymphatics of the deep inguinal group are also situated in the femoral triangle.

700 The obturator nerve supplies the adductor group of muscles. These are five in number: adductor longus, adductor brevis, adductor magnus, gracilis and obturator externus. As their names suggest, their principal function is to adduct the hip joint.

In general, they arise from the inferior ramus of the pubis and fan out to be inserted into the linea aspera of the femur. Gracilis, however, is inserted onto the upper medial aspect of the tibia and flexes the knee in addition to adducting the hip. Adductor magnus is really two muscles fused. It consists of an adductor portion (supplied by the obturator nerve), running from the pubis to the linea aspera, and a hamstring portion running from the ischial tuberosity to the adductor tubercle on the lower medial aspect of the femur. The strong medial collateral ligament of the knee joint is its phylogenetically degenerated distal portion that used to attach to the tibia, as it does in some animals. The hamstring portion is supplied by the tibial division of the sciatic nerve and extends the hip like all hamstrings.

Obturator externus arises from the outer aspect of the obturator membrane and adjacent pubis and ischium. Its tendon is inserted into the trochanteric fossa of the femur (a pit on the medial aspect of the greater trochanter). It rotates the hip laterally.

701 The adductor canal contains the femoral vessels, the saphenous nerve and the nerve to vastus medialis (both from the femoral nerve), and the terminal part of the obturator nerve. It conducts them from the apex of the femoral triangle superiorly, to the adductor hiatus in adductor magnus inferiorly. It is a triangular cleft between muscles situated under the middle third of sartorius, having medial, anterior and posterior walls which are formed from sartorius, vastus medialis and adductor longus, respectively, in the upper part. Inferiorly, adductor magnus forms the posterior wall.

B Applied answers

702 A finger in the femoral canal will be close in contact with the femoral vein laterally, the lacunar ligament and conjoint tendon medially and the inguinal ligament (Poupart) and spermatic cord anteriorly. Posteriorly, and deep to the femoral canal, lies the superior ramus of the pubis and pectineus muscle. Due to these relationships, operations for enlarging the femoral canal or ring must be made medially or anteriorly, where no important structure will be damaged.

703 To find the femoral vein, it is most useful to know that it lies medial and parallel to the femoral artery, just below the inguinal ligament. The femoral artery is easily located by finding its pulsation midway between the anterior superior iliac spine and the pubic symphysis. Both artery and vein lie within the femoral sheath, so the vein tends to be well anchored. With one finger on the arterial pulsation, the needle is introduced 0.5 cm medially. When actually attempting to obtain arterial blood, it is not uncommon to sample venous blood mistakenly, as the two vessels are so close together.

704 By observation, a femoral hernia lies below and lateral to the pubic tubercle, whereas the neck of an indirect inguinal hernia lies above the tubercle and passes either over or just medial to it.

705 The long saphenous vein is homologous to the cephalic vein. The short saphenous and basilic veins are also homologues.

706 The lymphatic drainage from the glans penis is to the superficial lymph nodes. This must not be confused with the lymphatic drainage of the testes, which is to the para-aortic nodes of the posterior abdominal wall.

707 A swelling within the psoas major muscle may cause pressure on parts of the lumbar plexus and the lumbosacral trunk (L4, 5) which is found within its bulk. It is also possible that a swollen psoas muscle might compress structures passing below the inguinal ligament, such as the femoral nerve and its branches. The tingling in the big toe can be easily explained by pressure on the L5 spinal nerve. Pressure on the femoral nerve, via its saphenous nerve component (L4), would give tingling in the medial aspect of the foot (the saphenous nerve is the terminal sensory branch of the femoral nerve). The numbness on the lateral part of the buttock and thigh might be due to compression of the lateral femoral cutaneous nerve (L2, 3) as it passes below the inguinal ligament, just inferior to the anterior superior iliac spine.

708 The long saphenous vein lies just anterior to the medial malleolus at the ankle, and crosses the subcutaneous tibia ('shinbone') in the lower leg. It is here that the saphenous nerve often entwines round the vein, so during the operation to strip out the vein, a portion of the nerve may easily be damaged.

709 Not only will the femoral artery and vein be endangered as they pass down the front of the thigh towards the subsartorial canal, but also will their branches and tributaries. Midway down the thigh, a knife piercing sartorius and the femoral vessels will also lacerate adductor longus, a fairly thin muscle, and damage the profunda femoris vessels on its deep surface. The injury is therefore very bloody due to damage to two large arteries and their accompanying veins.

C Information answers

710 The menisci are fibrous semilunar extensions of the knee joint capsule, centrally incomplete, attached to the intercondylar area of the tibia, a medial and lateral one existing in each joint

between the respective condyles of the femur and tibia. Although deepening the tibial articulation for the femoral condyles, due to their triangular cross-section, their main purpose is to partially compartmentalise the joint, therefore allowing rotation at the 'tibiomeniscal' part when the knee is flexed and extension/flexion at the upper 'meniscofemoral' part. This is an active rotation to allow change of direction at speed (performed by the differential pull of the hamstrings) compared to the passive rotation of the locking mechanism on knee hyperextension. The menisci also play a role in shock-absorbing the joint, when they bear some body weight on their compression during locking, with the tightening of the intra- and extracapsular ligaments which now take approximately half the body weight.

711 The 'carrying angle' between tibia and femur is due to the shorter distance between the centres of the knee joints compared to the femoral heads (the femoral shafts thus slope obliquely inwards). This gives a lateral component to quadriceps pull on the patella. Factors preventing lateral dislocation of the patella are:-
1. Bony – the forward prominence and larger height of the lateral femoral condyle with a larger lateral patellar facet to articulate with it.
2. Muscular – the horizontally orientated lower fibres of vastus medialis inserting into the medial aspect of the patella are indispensable to its stability.
3. Ligamentous – tension in the medial patellar retinacula.

712 There are about a dozen bursae in the region of the knee joint. Four directly communicate with the joint: the suprapatellar bursa, the popliteus bursa, the medial gastrocnemius bursa (always), and the lateral gastrocnemius bursa (usually). The semimembranosus bursa indirectly communicates via its opening into the medial gastrocnemius bursa. The suprapatellar (quadriceps) bursa extends one hand's breadth above the patella. Three further bursae are associated with the patella: the prepatellar bursa anterior to the bone, and the superficial and deep infrapatellar bursae on the respective sides of the ligamentum patellae. Other bursae are found around semimembranosus (between semi-

membranosus and the medial head of gastrocnemius) and around the collateral ligaments (one between the two parts of the tibial collateral ligament, one between tibial collateral ligament and pes anserinus, and one between the fibular ligament and biceps). Other bursae have been described.

713 The cruciate ligaments are two intracapsular ligaments situated in the knee joint. They cross each other at right angles, hence their name. They derive their respective names from their tibial attachments. Thus the anterior cruciate ligament is attached to the anterior intercondylar area of the tibia. It passes upwards, backwards and laterally, to be attached to the medial aspect of the lateral femoral condyle. The posterior ligament is attached to the posterior tibial intercondylar area. It passes upwards, forwards and medially to the lateral aspect of the medial femoral condyle. The ligaments prevent anterior and posterior displacement of the tibia on the femur; the anterior ligament preventing anterior movement, the posterior ligament posterior displacement.

In extension, the anterior ligament becomes taut. Further extension results in passive medial rotation of the femur on the tibia, causing all ligaments to tighten (including the posterior fibres of the posterior ligament) and anatomical locking of the knee. The anterior fibres of the posterior ligament become taut in flexion. Note that the clinical term 'locking' refers to the inability to extend the knee. This is commonly due to a foreign body, usually loose cartilage, in the joint.

714 The hamstrings consist of the semitendinosus, semimembranosus, the long head of biceps femoris and half of adductor magnus. They all arise from the ischial tuberosity, are supplied by the tibial division of the sciatic nerve and are situated on the posterior aspect of the thigh.

With the exception of adductor magnus, they are inserted into the upper aspect of the lower leg bones. Semitendinosus is inserted into the upper part of the subcutaneous surface of the tibia behind sartorius and gracilis. Semimembranosus is inserted via a broad aponeurosis to the posterior part of the knee joint capsule forming the oblique popliteal ligament, and over popliteus to the soleal line. Biceps is inserted into the head of the fibula.

Together, the hamstrings flex and actively rotate the knee while extending the hip.

Adductor magnus reaches only as far inferiorly as the adductor tubercle of the femur, with the medial collateral ligament of the knee joint being its phylogenetically degenerated tibial attachment. It cannot, therefore, have an action on the knee in humans, although in some animals it does.

715 The knee jerk involves stretching the ligamentum patellae by striking it with a hammer. This in turn stretches the muscle fibres of the quadriceps fermoris group (rectus femoris, vastus medialis, vastus intermedius, vastus lateralis). The stretch receptors in the muscles are excited and send sensory impulses to the spinal cord via the femoral nerve (L2, 3, 4). Motor impulses are sent back to the quadriceps via the same nerve, causing the muscle to contract. The knee jerk depends on the integrity of the quadriceps muscle and ligamentum patellae, both sensory and motor limbs of the femoral nerve and the spinal cord in segments L2–L4.

716 The popliteal fossa is situated behind the knee joint. Its floor may be divided into thirds. From superior to inferior, this is formed by the popliteal surface of the femur, the capsule of the knee joint and the fascia covering popliteus. Its roof is the skin and superficial and deep fasciae of this portion of the lower limb. Its boundaries are: laterally, biceps femoris and the lateral head of gastrocnemius with plantaris (from top to bottom); medially, semimembranosus and semitendinosus above and the medial head of gastrocnemius below. It contains connective tissue, lymph nodes, the popliteal artery and vein, and the terminal branches of the sciatic nerve. The popliteal artery is the continuation of the femoral artery as it passes through the adductor hiatus. It is the deepest structure in the popliteal fossa. It is accompanied by the popliteal vein. The short saphenous vein pierces the deep fascial roof of the fossa to enter the popliteal vein. The sciatic nerve bifurcates under biceps into the common peroneal and tibial nerves. The former runs along biceps across the lateral aspect of the fossa. The latter runs superficial to the popliteal vessels in the fossa's median plane.

717 Popliteus arises by muscular fibres from the upper medial aspect of the tibia above the soleal line. It passes upwards and laterally, to be inserted via a tendon into a depression just below the lateral epicondyle of the femur. The tendon lies within the capsule of the knee. The popliteus medially rotates the tibia on the femur. During extension, the ligaments of the knee become taut by lateral rotation of the tibia on the femur. This 'locks' the knee. For flexion to occur, 'unlocking' must first take place. This action is performed by popliteus, which medially rotates the tibia on the femur.

C Applied answers

718 The lowest fibres of vastus medialis waste rapidly after an effusion into the knee joint or prolonged bed rest. The knee now feels unsteady because:
   1. Lateral deviation of the patella gives abnormal afferent proprioceptive impulses (the nerve to vastus medialis conveys most of the proprioception from the knee).
   2. The last 20° of extension are much weakened; since this is an essential function of these lowest fibres.

719 The patella is the largest sesamoid bone in the body. It develops in hyaline cartilage and ossifies frequently from more than one centre between 3 and 6 years. Usually these coalesce, but sometimes the superolateral and other centres may ossify independently. The bipartite patella may cause misinterpretation as a fracture to an unwary casualty officer! Although usually asymptomatic, abnormal patella tracking can give knee pain at onset of athletic activity. A direct blow may fracture a bi- or tripartate patella into its constituent parts. Transverse fractures by the sudden pull of quadriceps (e.g. slipping and attempting to prevent a backward fall) occur more often in persons with unfused or poorly fused ossification centres. It is useful to compare both sides if in doubt over X-ray interpretation, as anomalies of ossification are usually bilateral.

720 The extensive superior communication between the suprapatella bursa and the knee joint enables a stab wound above the knee to result in a septic arthritis or foreign body being introduced into the

joint, while a fracture of the distal third of the femur may cause a haemarthrosis of the knee joint which will require drainage. Also, an effusion can be drained away from the joint by aspiration of the suprapatellar bursa.

721 Muscular factors are of prime importance in knee joint stability. Many sports injuries are preventable through appropriate conditioning of quadriceps femoris (especially its inferior horizontal fibres). One of the functions of the patella is to articulate with the femur giving anterior support to it when weight-bearing on a flexed knee. The joint functions surprisingly well after ligament strain if quadriceps femoris is well developed. Ligament injuries occur from any blow forcing the knee joint to move in an abnormal plane. Cruciate ligament injuries occur with excessive movement in the anteroposterior plane of the flexed knee.

722 Pain in the knee joint may in fact be due to referred pain from the hip joint. The nerve supply of the hip joint is usually from the femoral nerve and branches from the obturator, with occasional contributions from the nerve to quadratus femoris and the superior gluteal nerve. The knee has a very similar supply, with the femoral, obturator and sciatic nerves all contributing some twigs to the joint. It is easy to see how pain from the two joints is easily confused by the brain. Hilton's law states that the motor nerve to a muscle tends to give a branch of supply to the joint which the muscle moves, and another branch to the skin over that joint.

723 The medial meniscus is more commonly torn because it is attached directly to the medial collateral ligament. When a twisting injury affects the knee joint, tension is applied to the collateral ligaments. If the meniscus is held taut by the medial ligament and the bony condyles continue to move, then a tear may appear in the meniscus. 'Bucket handle' tears of the medial meniscus are quite common and are usually felt as pain along the joint line itself.

724 In extension, the patella is anterior to the knee joint and femur, but is separated from them by the synovial membrane and suprapatellar bursa. A knee joint distended with an effusion results in

fluid lying in this intervening space, and when pressed down sharply the patella can be bounced up and down on this fluid cushion. The fluid often tracks to the suprapatellar pouch, or bursa, deep to the quadriceps tendon, and may lie some 5 cm superior to the patella.

725 Swellings of the bursae around the knee are named after certain occupations which are supposed to cause trauma, and hence effusions, of these bursae. Housemaid's knee is a swelling of the prepatellar bursa, which lies anterior to the lower part of the patella and upper ligamentum patellae. This swelling is said to be caused by scrubbing floors on hands and knees. Clergyman's knee is said to be due to excessive kneeling at prayer! It is an effusion in one of the infrapatellar bursae which lie between the tibia and ligamentum patellae and the skin.

726 The patella is the largest sesamoid bone and ossifies around the age of 3 years. Women dislocate their patellae more commonly than men, probably due to their wider pelvis and shorter structure, and therefore increased lateral pull of the quadriceps. Factors which prevent lateral patellar dislocation include the horizontal direction of the lower fibres of vastus medialis, the size and shape of the lateral femoral condyle and the shape of the patella itself. The last two features are readily seen on a skyline view X-ray of the knee.

727 The patella is a sesamoid bone which develops in hyaline cartilage and begins to ossify from several centres at about 3 years. The superolateral angle may ossify independently. These congenital bipartite patellae are easily confused with a fracture.

728 The ossification centres of the distal end of the femur and proximal end of the tibia are used in determining a baby's age. The centre of ossification for the distal end of the femur usually appears before birth, in both sexes, whereas the proximal tibial centre appears at birth. The presence or absence of these two epiphyses on an X-ray can give a fairly good idea of the age of the baby at the time of death.

729 A torn meniscus, which is more commonly the medial, may result in a loose segment of the cartilage moving independently within the joint.

This free-moving cartilage may become wedged between the articular surfaces of tibia and femur, making further extension impossible and thus 'locking' the knee. A minor form of this condition may cause a 'clicking' knee joint.

730 The popliteal pulse is extremely difficult to palpate in most people. This is because it is the deepest-lying structure in the popliteal fossa. Having pierced the adductor magnus at the adductor hiatus, it lies deep to its vein, which is in turn, deep to the tibial nerve in the popliteal fossa. All these structures are embedded in fat, and both the artery and the vein are additionally surrounded by a tough fibrous sheath. Flexion of the knee often facilitates palpation due to relaxation of nerve and sheath.

731 All the structures passing through the popliteal fossa may cause a swelling in this region. A popliteal aneurysm and popliteal or short saphenous varicosities are the most likely causes. However, neuromata of the tibial, sural or common peroneal nerves can occur as well as effusions in the semimembranosus bursae, or a cystic swelling originating from the joint capsule itself (Baker's cyst).

732 The three tendons most easily palpable, and often visible at the back of the knee in flexion, are biceps femoris laterally, and semitendinosus and semimembranosus medially.

733 This is a typical 'bumper-bar' injury, which commonly results in a fractured neck of the fibula. At this site, the common peroneal nerve (common fibular is the new name for this nerve – Nomina Anatomica 1989 – since the word peroneus has been replaced by fibularis, and peroneal by fibular) winds into the lateral, and later the anterior, compartment of the lower leg, and so is often damaged. The branches of the common peroneal nerve are the superficial and deep peroneal nerves. The deep nerve runs with the anterior tibial artery to supply the extensor group of muscles, including tibialis anterior, whereas the superficial nerve supplies the peroneal muscles which are the major evertors of the foot. Damage to the deep peroneal nerve results, therefore, in an inability to dorsiflex the foot, whilst an injured superficial peroneal nerve results in an inverted foot. A combination of

these is 'foot-drop' which is thus seen as an inverted, plantar flexed foot. This results in the patient catching his big toe on the swing-through of the leg when walking, and scuffing the anterolateral part of this shoes when he walks.

D Information answers

734 The interosseous membrane running between fibula and tibia separates the leg into an anterior (extensor) compartment and a posterior (flexor) compartment. The peroneal muscles, attached to lateral aspect of the fibula, form a third, lateral, compartment. Each compartment is served by its own nerve – viz. anterior: deep peroneal nerve; posterior: tibial nerve; lateral: superficial peroneal nerve. The anterior compartment consists of tibialis anterior, extensor hallucis longus, extensor digitorum longus and peroneus tertius. The posterior compartment may be subdivided into superficial and deep parts. The superficial part comprises gastrocnemius, plantaris and soleus. The deep part consists of popliteus, flexor digitorum longus, flexor hallucis longus and tibialis posterior. The two parts are separated by a fascial septum – the deep transverse fascia of the leg. The lateral compartment contains peroneus longus and peroneus brevis.

735 The principal plantar flexors of the ankle are gastrocnemius and soleus, which insert onto the calcaneus. In theory, the muscles passing behind the flexor and peroneal retinacula should also be plantar flexors – viz. peroneus longus, peroneus brevis, tibialis posterior, flexor digitorum longus and flexor hallucis longus. Their mechanical advantage is poor, however, and cutting the tendo calcaneus results in severe weakness of this movement.

Dorsiflexion is performed by those muscles passing anterior to the joint – viz. tibialis anterior, extensor hallucis longus, extensor digitorum and, its morphologically emerging offshoot, peroneus tertius (falls short of the fifth phalanx).

736 The tissues of the lower limb, superficial to the deep fascia, are drained of venous blood by the great and short saphenous veins. Both arise from the venous network on the dorsum of the foot.

The long saphenous vein arises from the medial aspect of this network. It ascends, passing immediately anterior to the medial malleolus. It continues upwards and backwards on the medial aspect of the leg, in company with the saphenous nerve, to pass a hand's breadth behind the medial side of the patella. It curves forwards as it ascends the thigh to reach the saphenous opening 2.5 cm below the inguinal ligament, where it pierces the deep fascia and enters the femoral vein. Various perforating veins run between the long saphenous vein and the deep veins of the calf, being arranged in three groups related to the adductor canal, the calf muscles and a collection just proximal to the ankle joint. The short saphenous vein arises from the lateral aspect of the venous network of the dorsum of the foot. It ascends immediately behind the lateral malleolus and passes upwards and posteriorly to reach the middle of the back of the leg. It runs upwards in company with the sural nerve to the popliteal fossa, where it pierces the deep fascia and enters the popliteal vein.

737 At the ankle, each tendon is contained within a synovial sheath which lubricates it as it moves under a retinaculum. The retinaculum ties the tendon down and prevents 'bowstringing'. The inferior extensor retinaculum consists of superficial and deep layers. The tendons run between these layers in their synovial sheaths, separated from their neighbours by septa which join the superficial and deep retinacular parts. From medial to lateral, these extensor tendons are tibialis anterior, extensor hallucis longus, extensor digitorum longus and peroneus tertius.

738 The peronei are so named because they take origin from the fibula (L. *fibula* = pin; Gk. *peroneus* = skewer). Peroneus longus originates from the upper two-thirds of the fibula's lateral surface, peroneus brevis from the lower two-thirds. Both tendons run behind the lateral malleolus and are held in position by the peroneal retinaculum. The tendons are separated by the peroneal tubercle of the calcaneus – the brevis passing above it, the longus below, and both being held in position by the inferior peroneal retinaculum. Peroneus brevis is inserted into the base of the fifth metatarsal. Peroneus longus enters a groove in the cuboid and crosses deep in the sole

of the foot, to be inserted into the base of the first metatarsal and adjacent medial cuneiform. Both are supplied by the superficial branch of the common peroneal nerve. Both muscles plantar flex the ankle. They are also evertors, the movement occurring at the tarsal joints. This is their unique action, no other muscle being capable of this. Peroneus longus helps support the transverse arch of the foot as it crosses the sole. Peroneus brevis maintains the lateral longitudinal arch.

739 Around the ankle joint, the deep fascia is thickened in various regions to form a series of retinacula, which bind the tendons down and keep them in position. The extensor tendons are held in place by two extensor retinacula – superior and inferior. The superior is attached to the anterior aspects of the tibia and fibula. It is split to enclose tibialis anterior. The inferior extensor retinaculum is the shape of a Y lying on its side. The base is attached to the upper anterior aspect of the calcaneus. The superior limb is attached to the medial malleolus, the inferior to the plantar fascia. The tendons of tibialis anterior, extensor hallucis longus, extensor digitorum and peroneus tertius are held in place by the extensor retinaculum. The anterior tibial vessels and deep peroneal nerve also run under it. The flexor retinaculum extends from the medial malleolus to the medial aspect of the calcaneus. Tibialis posterior, flexor digitorum longus, the posterior tibial vessels and tibial nerve run deep to this structure. The peroneal muscles are held in pace by two peroneal retinacula – superior and inferior. The superior encompasses both the longus and brevis tendons and runs from the lateral malleolus to the calcaneus. The inferior retinaculum is attached to the peroneal tubercle of the calcaneus which separates the two tendons (brevis being superior). It is attached above and below to the calcaneus.

740 The ankle joint is a synovial joint of the hinge variety. Being a hinge, it has strong collateral ligaments to prevent movements other than in the anteroposterior plane. The medial ligament (deltoid ligament) is triangular. Its apex is attached to the medial malleolus. Its base consists of a superficial and a deep part. The deep fibres are attached to the talus, the superficial fibres to the navicular (anteriorly), calcaneus (centrally) and talus (pos-

teriorly). The lateral ligament consists of three distinct parts: the anterior talofibular ligament (anteriorly), running from lateral malleolus to the lateral surface of the talus; the calcaneofibular ligament (centrally), running from lateral malleolus to calcaneus, and the posterior talofibular ligament (posteriorly), running from the medial aspect of the lateral malleolus to the posterior tubercle of the talus.

741 The ankle jerk is elicited by tapping the tendo calcaneus (Achilles) with a hammer. This stretches the muscles attached to it (i.e. gastrocnemius and soleus). Afferent impulses pass via sensory fibres in the tibial nerve, which supplies these muscles, to cord segments L5, S1 and 2. Efferent impulses travel from these segments via motor fibres in the tibial nerve to these same muscles, causing them to contract. The presence of the ankle jerk depends therefore upon the integrity of the gastrocnemius and soleus, the tendo calcaneus, the sensory and motor limbs of the tibial nerve, and cord segments L5, S1 and 2.

D Applied answers

742 Causes of acute calf pain during activity include a tear of the gastrocnemius, commonly seen in racquet sports. An Achilles tendon rupture may present as an acute tearing pain in the heel during athletics, as if having been kicked or struck on the heel from behind. The clinical importance of the plantaris muscle lies in the possibility of its rupture during sudden violent ankle dorsiflexion, snapping the long, slender tendon of this feeble atavistic muscle (most cases are accompanied by tears of triceps surae). This injury is common in ballet dancers, sprinters and basketball players with the pain being so severe that weight-bearing is almost impossible!

743 'Shin splints' is the painful condition of the anterior leg after rigorous or lengthy exercise – often people who lead sedentary lives develop pains here on long walks. The anterior tibial muscles swell, causing increased pressure in the anterior compartment of the leg. This may compromise blood flow to these muscles (a mild form of compartment syndrome). The condition may also

occur in trained athletes who don't warm up adequately.

744 Gastrocnemius originates from the femoral condyles, whereas soleus originates from the fibula and soleal line of the tibia and does not cross the knee joint. Both insert together as the tendo calcaneus. In order to demonstrate them separately, one must first flex the knee to neutralise gastrocnemius and then resist plantar flexion. This will single out soleus, whereas with the knee extended both will contract in plantar flexion.

745 The short saphenous vein and branches of the sural nerve may be in very close contact in the calf, so stripping out the vein may traumatise the sural nerve, which supplies to skin of the lateral side of the foot and little toe. For the same reason, stripping the long saphenous vein may leave tingling along the medial side of the foot due to damage to the saphenous nerve.

746 In most cases it is the deep venous system which is investigated. This is achieved by injection of contrast medium into a superficial vein on the dorsum of the foot. Constrictor bands are placed at the ankle and below the knee which occlude the superficial system. Occlusion of the superficial veins causes the deep venous system to fill. Indications for this procedure include diagnosis of deep vein thrombosis, and the location of varicosities in the perforating veins which pass from superficial to deep venous systems. Ultrasound is another useful diagnostic procedure.

747 The most constant site of the long saphenous vein which, prior to the popularisation of central venous catheterisation, was commonly used for 'cut-downs', is at the ankle. Here the vein runs 2–3 cm anterior to the medial malleolus, and can be seen or palpated in most normal individuals.

748 The tendo calcaneus is the strongest tendon in the body and ruptures only if violent force is used. Usually this happens to athletes or entertainers who carry on their sport beyond their prime, though the popularity of such sports as squash and rock climbing has produced quite a few young people with this injury. The injury is disabling, as any

useful plantar flexion proves impossible. Many surgeons advise operative suture as early as possible.

749 Normally, most people have both a posterior tibial pulse (felt just behind the tendon of flexor digitorum longus at the medial malleolus) and a dorsalis pedis pulse (felt on the dorsum of the foot between the tendons of extensor hallucis longus and extensor digitorum longus). Usually the dorsalis pedis artery is a direct continuation of the anterior tibial artery, but occasionally the arterial supply of the foot is from a perforating branch of the peroneal artery and, in these cases, the dorsalis pedis pulsation is very weak or even absent. Absence of either posterior tibial or dorsalis pedis pulsation occurs in about 5% of the normal population.

750 The ankle joint is most unstable with the foot plantar flexed, which happens when one is standing on one's toes or walking downstairs, or walking on the level with high heels! However, in dorsiflexion it is a rigid stable joint. This is due to the shape of the talus, which is broader anteriorly. In dorsiflexion, the talus is forced into the tibiofibular socket, pushing these bones slightly apart and causing the malleoli to grip the talus firmly. This produces a very firm platform. The fibula rotates and a small gliding movement occurs at the superior tibiofibular joint.

751 A torn deltoid (medial) ligament is a serious injury for two reasons. Firstly, it causes instability of the ankle joint because this ligament maintains the mortice effect of the tibia and fibula with the talus, by connecting the medial malleolus to the sustentaculum tali and tuberosity of the navicular bone. Secondly, it attaches to the plantar calcaneonavicular (spring) ligament, which is important in the maintenance of the medial longitudinal arch of the foot.

752 In 1768, Percival Pott described a lateral outward fracture-dislocation of the ankle. This injury has retained his name, and is divided into three degrees of severity. The first degree involves a fracture of the lateral malleolus alone, the fracture line running obliquely upwards and backwards. The second degree involves the talus moving laterally, causing

an oblique lateral malleolar fracture, with either a transverse fracture of the medial malleolus or severe disruption of the deltoid ligament. The third degree cases are due to continued rotational forces on the talus, forcing it laterally and backwards. This produces fractures of both malleoli and of the posterior portion of the tibial articular surface. A third degree fracture thus leaves a most unstable ankle, often with three separate fracture lines. Internal fixation is required to restore the original position of the mortice joint into which the talus fits.

E Information answers

753 'Extra' foot bones include the os trigonum, occurring when the lateral tubercle of the posterior process of the talus fails to unite with the body during ossification, and the os vesalianum pedis, which occasionally appears near the base of the fifth metatarsal. This can be confused with avulsion of the fifth metatarsal tuberosity, and so can a secondary ossification centre in children and adolescents of the fifth metatarsal tuberosity. If doubt exists, the other foot should be X-rayed for comparison as anomalies of ossification are usually bilateral.

754 The four layers of plantar muscles and tendons are specialised to maintain the arches of the foot, enabling efficient weight-bearing, propulsion and standing on uneven ground. Consequently, these muscles have gross functions and several have names implying delicate individual functions they rarely, or are unable to, perform. The first layer includes three short muscles connecting the calcaneus to the phalanges – abductor hallucis brevis, flexor digitorum brevis and adbuctor digiti minimi. The second layer has the tendons of flexor digitorum longus connected with flexor accessorius and the lumbricals, and the tendon of flexor hallucis longus giving 'beef to the heel' deep to these. The third layer, in the anterior half of the sole, contains the shorter muscles of great and smallest toes, namely adductor hallucis, flexor hallucis brevis and flexor digiti minimi brevis. The fourth layer contains the interossei and the long tendons of peroneus longus and tibialis posterior, crossing the sole to reach their insertions, namely the first

metatarsal/medial cuneiform and navicular, respectively.

755 The ability to invert and evert the foot enables walking on uneven surfaces and the ability to change direction at speed (cf. menisci and active knee rotation). Inversion, with forefoot adduction by tibialis anterior and posterior, contrasts with eversion and forefoot abduction by peroneus longus, brevis and tertius (in each case the latter muscle counteracts the unwanted plantar flexion of the former). As all attach to the forefoot anterior to the midtarsal joint (calcaneocuboid joint and talonavicular part of the talonaviculocalcanean joint), initially the ligaments of these joints (calcaneocuboid ligament and lateral limb of bifurcate ligament) are 'wound up' soon after minimal movement here, with the rotatory tension transferred via them to the subtalar joint (an orthopaedic term covering the talocalcanear part of the talocalcaneonavicular joint anteriorly, and the talocalcanear joint posteriorly). Most of the movements occur here, beneath the talus, along an axis through the centres of curvature of the upward-facing, concave anterior and convex posterior, articulating facets of the calcaneus. A marvellous example of bioengineering is seen as the long tendons pull at right angles to this superolateral axis to maximise their advantage, while the strong cervical ligament (connecting the talus with calcaneus in the tarsal sinus) and the calcaneofibular part of the lateral ligament complex of the ankle joint both lie parallel to this axis, giving minimum hinderance to these movements. Extrapyramidal rotation of the flexed knee, or if extended, of the hip joint, keeps the body facing in the same direction if required, counteracting the accompanying movements of the forefoot (similar, elbow and shoulder joint movements accompany pronation and supination of the forearm).

756 The anatomical structures creating the imprint of a wet foot on the bathroom floor include the tuber calcanei and the overlying fibrofatty tissue superficial to the inferior third of its posterior surface (the heel!), all the constituents of the lateral longitudinal arch, namely the calcaneus, cuboid and lateral two metatarsal heads (the cuboid bears weight the least), the pillars of the medial longitudinal arch – the tuberosity of calcaneus and the

heads of the medial three metatarsals – and the pads of the distal phalanges.

757 The factors that maintain the integrity of the arches of the foot are mainly ligamentous for static body weight, and muscular for stabilisation against the huge propulsive thrusts tending to flatten the arches in walking and running. No bony factor is responsible for the stability of the longitudinal arches although bony factors play a minor role in the support of the transverse arch. For the medial arch, the interosseous ligaments (especially between the talus and calcaneus), the plantar aponeurosis, stretching like a tiebeam or bowstring between the arch pillars, and the spring ligament, supporting the head of talus between navicular and calcaneus, are all essential. Muscles, although inactive when standing still, play a vital role in arch support during movement. The tendon of flexor hallucis longus ties the undersurface of the sustentaculum tali to the medial three toes (it gives slips to the second and third toe tendons of flexor digitorum longus). The short muscles of the first plantar layer inserting into the medial three toes (abductor hallucis and the medial half of flexor digitorum brevis) likewise assist. Tibialis anterior and posterior have significant actions, but are less important factors in raising the medial border of the foot from the ground. Ligaments play a more important role in lateral arch stability; the bowstringing effect of the lateral aspect of the plantar aponeurosis is reinforced by the long and short plantar ligaments. The tendon of peroneus longus crucially suspends this arch as it enters the subcuboid groove in the fourth plantar layer, with strong assistance from the bowstringing flexor digitorum longus tendons to the fourth and fifth digits, aided by flexor accessorius. The transverse arch's minor bony supporting factor is the lateral cuneiform bones resting on the cuboid with minimal pressure. The intermediate cuneiform is, however, wedge-shaped the wrong way for arch stability! The interosseous ligaments more importantly tie the bases of the five metatarsals and the cuneiforms together, with the assistance of the transverse plantar ligament. Peroneus longus bowstrings the transverse (hemi)arch, between the cuboid and navicular bones.

758 The four named nerves which supply sensation to the dorsum of the foot are the saphenous nerve, the sural nerve, the deep peroneal nerve and the superficial peroneal nerve. The saphenous nerve – a branch of the femoral nerve – supplies skin on the medial side of the leg and the adjacent area of the dorsum of the foot as far as the metatarsophalangeal joint of the hallux. The sural nerve – a branch of the tibial nerve – runs behind the lateral malleolus onto the lateral side of the foot, supplying it on dorsal and plantar aspects. It also supplies the little toe. The deep peroneal nerve supplies, via its medial terminal branch, a wedge of skin on the dorsum of the foot on adjacent sides of the hallux and second toes. The remainder of the dorsum is supplied by the superficial peroneal nerve via its medial and lateral terminal branches.

759 The arches of the foot are the medial and lateral longitudinal arches and the transverse arch. The last is only a half-arch, being deficient medially. Various factors are involved in their maintenance. Firstly, the bones are wedge-shaped, tapering inferiorly. Secondly, the bones are bound together by ligaments. The ligaments on the plantar aspect are the stronger. Thirdly, the ends of the arches are connected by muscles and tendons like a bowstring. Fourthly, the arches are suspended from above by long tendons; viz. peroneus longus, tibialis posterior and tibialis anterior. The medial longitudinal arch consists of calcaneus, talus, navicular, medial cuneiform and first metatarsal. The inferior aspects of these bones are connected by the calcaneonavicular ligament. The plantar aponeurosis, flexor hallucis longus, abductor hallucis and the medial portion of the flexor digitorum muscles connect the two ends of the arch. Tibialis anterior and posterior suspend the navicular from above. The lateral arch consists of calcaneus, cuboid and fifth metatarsal. They are tied together by long and short plantar ligaments inferiorly. The ends of the arch are joined by the plantar aponeurosis, abductor digiti minimi and the lateral portion of the flexor digitorum muscles. The peroneal muscles suspend it from above. The transverse arch is between the cuneiform (wedge-shaped) bones and the metatarsal bases, held together by the deep transverse plantar and interosseous ligaments, and are suspended by peroneus longus and tibialis posterior.

760 The line of gravity passes through the body of the axis, anterior to the thoracic spine and sacrum. It then passes behind the hip joints but anterior to the knee and ankle joints.

761 In walking, one lower limb supports the body weight while the other is swinging forwards. The roles of supporting and swinging are then reversed.

Let us consider the right leg to be swinging and the left leg to be supporting initially. As the right foot is taken off the ground the pelvis is tilted by the left abductors (gluteus medius and minimus), so that the right ilium is raised. This allows the foot to clear the ground during swinging. The right hip flexes, swinging the right leg forwards, first by gravity and then by iliopsoas. The right knee is also flexed, by gravity and the hamstrings, and the foot is dorsiflexed by the extensors and tibialis anterior. This flexion of the knee and extension of the ankle is further increased to assist clearance of the limb from the ground during the swinging phase. As the limb is moved anterior to the axis of the body, the knee is extended by the quadriceps. The right ilium is thrust forward and the right leg rotated laterally to keep it in line with the direction of movement. The ankle is plantar flexed as the foot approaches the ground, with tibialis anterior 'paying out rope' to prevent foot-slap. As the right foot touches the ground, the supporting phase commences. The left hip abductors are relaxed and the pelvis returns to its level attitude. The body weight is moved anteriorly from the heel along the outer side of the sole of the foot to the metatarsal heads. At the same time, the body weight is brought above the right foot by extension of the hip (gluteus maximus and hamstrings). The knee is kept extended and the foot dorsiflexes as the body moves over it.

As the body weight moves anterior to the position of the right foot, the foot is plantar flexed (gastrocnemius and soleus) and the heel leaves the ground. The metatarsophalangeal joints are flexed as a consequence. The flexor muscles of the toes (particularly flexor hallucis longus) now contract and push the body weight forwards to take the next step. The right foot leaves the ground and the process is repeated.

762 Lymph from the superficial tissues of the heel will run in the lymphatics following one or other of the

saphenous veins. That following the long saphenous vein will eventually drain to the vertically disposed groups of superficial inguinal nodes around the termination of the vein. The lymph vessels following the short saphenous vein will pierce the deep fascia in the popliteal fossa and enter the popliteal nodes. From here lymph passes to the deep inguinal nodes around the femoral vein in the femoral triangle.

E Applied answers

763 Common foot fractures include avulsion of the tuberosity of the fifth metatarsal by the peroneus brevis tendon during sudden violent inversion of the foot particularly with the foot plantar flexed (walking down stairs, or wearing high heels). To the unwary, some children and adolescents may have secondary ossification centres for the lateral surfaces of the fifth metatarsal tuberosities. To distinguish these from a fracture, X-rays of both feet may be needed in this age group. Persons who fall from great heights onto their heels often fracture their calcanei, usually into several fragments – a very disabling condition if the subtalar joints are disrupted as inversion and eversion are compromised. Because the second metatarsal bone has little movement at its distal joint, it is particularly liable to fracture, e.g. the 'march fracture' in those who are out of condition performing strenuous exercise programmes, marathons etc. A fractured neck of talus can occur in severe dorsiflexion of the ankle, e.g. pressing hard on the brake pedal in a head-on collision.

764 The arteries in the sole of the foot are derived from the posterior tibial artery which divides deep to abductor hallucis. The medial plantar artery passes distally between this muscle and flexor digitorum brevis. The larger lateral plantar artery runs forward between flexor digitorum brevis and accessorius. At the base of the fifth metatarsal it forms the plantar arterial arch, completed medially by the deep plantar branch of the dorsalis pedis artery. Wounds of this arch can result in severe bleeding and its control is difficult owing to its depth and related structures.

765 In infants the flat appearance of feet is normal, a result of the subcutaneous fat pads in the soles. Bony arches are present at birth but only become visible after the baby has walked for a few months. Flat feet in adolescents and adults are caused by 'fallen arches', usually the medial longitudinal arch. During long periods of standing, older persons, or those who rapidly gain weight, stretch the plantar ligaments and aponeurosis, which are nonelastic structures, with the loss of their important bowstringing effect in the stationary foot. The strain on the spring ligament eventually makes it unable to support the head of the talus, and, as a result, flattening of medial longitudinal arch occurs with lateral deviation of the forefoot. Fallen arches are painful owing to stretching of the plantar muscles and strained plantar ligaments. Pes planus is a true flattening from osteomuscular causes (see Question 768).

766 In severe inversion injuries, tension is exerted on the laterally placed muscles and, in particular, the peronei. Peroneus brevis, originating from the fibula, inserts into the tuberosity of the fifth metatarsal. Tension on this insertion often results in its fracture by avulsion. However, a fracture should not be confused with an epiphysis, which is frequently present at this site.

767 Talipes equinovarus is a deformity of the foot consisting of the hindfoot being drawn up by tight tendo calcaneus (L. *equino* = horse-like), the sole facing inwards (varus), as well as a slight adduction of the forefoot. The head of the talus may be felt anterior to the lateral malleolus instead of in its normal position below and anterior to the medial malleolus. This new position allows the anterior compartment (tibial) muscles to draw the foot into further adduction. It may be a congenital deformity, for which intrauterine pressure has been blamed, or it may be found secondarily to conditions such as spina bifida.

768 Pes planus is true flattening of the longitudinal arch, caused by abnormalities of the talonavicular joint, naviculocuneiform joint or first cuneiform metatarsal joints. These may be of bony origin or of muscular origin, the latter being due to inefficiency of the tibialis posterior tendon in maintaining the arch.

769 Pes cavus is an abnormal elevation of the longitudinal arch, which is seen in numerous neurological disorders such as poliomyelitis, spina bifida or Friedreich's ataxia. It is due to shortening of the plantar muscles and fascia. In severe cases, lack of interossei muscle function may result in severe clawing, which eventually leads to dorsal subluxation at the metatarsophalangeal joints. Occasionally, an extremely high arch may require removal of a bony wedge from the talus for its correction.

770 Lumbar sympathectomy will remove the sympathetic outflow to the arteries of the lower limb, and thus cause dilation of the collateral circulation in a leg with occlusive arterial disease of the main vessels. This operation is also sometimes carried out to benefit patients with chronic severe chilblains, by increasing the blood flow to the cold, blotchy feet. It is similar in rationale to thoracocervical sympathectomy for Raynaud's disease of the hands.

BACK

A Information answers

771 A typical cervical vertebra is slender in structure. Its body is wider from side to side than anteroposteriorly and possesses small synovial joints laterally. It has a large triangular vertebral foramen and small spines, which are bifid. The transverse process has a hole in it, the vertebrarterial foramen (old name is foramen transversarium), for the vertebral vessels. No other vertebra possesses these structures.

A typical thoracic vertebra possesses two facets on each side of the body, and one on its transverse process for the articulation of the ribs. No other vertebra has these facets. Its body is heart-shaped and the long spines slope inferiorly. The articular facets are in the vertical plane.

The lumbar vertebrae are relatively much more massive, with a large kidney-shaped body. The spines are quadrangular in shape and face directly backwards. The articular facets are arranged in the sagittal plane.

772 The spinous process of the seventh cervical vertebra is the most easily palpated, being much longer

than the others. For this reason it is called the vertebra prominens. However, just as often, that of T1 is the longest, so this feature cannot be relied upon for accurate vertebral identification by palpation.

773 During development the body of a vertebra is formed from a central mass – the centrum – and the two lateral parts form the neural arch. The centrum of the atlas is fused to the centrum of the axis, forming the dens (odontoid peg). The atlas is therefore missing the bulk of its body.

774 The iliac crests lie on a transverse plane through the level of the fourth lumbar vertebra. This is the level of the bifurcation of the abdominal aorta. Clinically, this plane is commonly used as a surface marking for lumbar punctures.

775 The direction of movement allowed between adjacent vertebrae depends upon the shape and direction of the articular processes. In the lumbar region the sagittal disposition of these allows flexion and extension more readily than other movements.

776 Vertebral bodies are joined by secondary cartilaginous joints. Each vertebral body is covered at its articular end with hyaline cartilage. Between the hyaline cartilage covering adjacent vertebral bodies is a fibrous disc. These joints are strengthened by the anterior and posterior longitudinal ligaments.

Between the lateral aspects of the bodies of adjacent cervical vertebrae there are, on each side, in addition to the central secondary cartilaginous joints, a small synovial joint (neurocentral/uncovertebral joints), although there is some debate as to whether these areas just represent lateral disc degeneration.

777 The specialised ligaments of the atlantoaxial joint are the transverse ligament of the atlas and the two alar ligaments.

The transverse ligament of the atlas unites the lateral masses of the atlas and runs posterior to the dens of the axis. It holds the dens firmly against the anterior arch of the atlas like a strap, preventing its posterior displacement towards the spinal cord but allowing rotation.

The alar ligaments, one on each side, arise from the top of the dens and run superolaterally to the medial sides of the occipital condyles. They hold the skull to the atlantoaxial unit, and limit rotation of the atlantoaxial joint.

778 The ligamentum nuchae is found in the cervical region, and is the continuation of the interspinous and supraspinous ligaments of the thoracic and lumbar vertebrae. It is attached to the external occipital crest and spines of the cervical vertebrae.

In quadrupedal mammals with large protuberant necks (e.g. horses) it is well developed, contains abundant elastic tissue and helps support the head. Man stands vertically, supporting his head on the cervical vertebrae, and consequently the human ligamentum nuchae is poorly developed.

779 The bodies of the vertebrae are united by the anterior and posterior longitudinal ligaments. The anterior longitudinal ligament is attached to the anterior surfaces of the bodies and intervertebral discs. The posterior longitudinal ligament lies in the vertebral canal and is attached to the posterior surfaces of the discs and margins of the bodies but not to the central portion of the bodies.

The spines are joined by interspinous ligaments, which run between adjacent spines, and the supraspinous ligaments, which are attached to the tips of the spinous processes. The ligamenta flava join the laminae of adjacent vertebrae and with them form the posterior wall of the vertebral canal. They contain an abundance of yellow elastic tissue, hence their name. (L. *flavus* = yellow).

A Applied answers

780 Kyphosis is a hump back (Gk. *kyphos* = humpbacked) and is used to describe an abnormal spinal curvature with a backward convexity. Scoliosis comes from the Greek for crookedness, and is a lateral curvature of the spine. Lordosis is a hollow back. A degree of lumbar lordosis is found in all normal healthy people.

781 Spina bifida is a failure of closure of the neural groove with associated abnormalities of the related

mesoderm, resulting in an incomplete bony neural arch. Most often found in the lumbosacral region, this condition is fairly common, especially in Wales and north-west England, and although no specific biochemical or chromosomal defect has been identified, it certainly has a familial tendency. Cytoplasmic inheritance has been suggested to explain the frequency amongst siblings. Different degrees of spina bifida range from complete non-fusion, myelocele through meningomyelocele, and meningocele, to spina bifida occulta, where only the bony canal is abnormal. Spina bifida occulta may be suggested by a thick tuft of hair in the lumbar region overlying the defect. These malformations of the spinal cord may result in incontinence and paralysis of lower limbs.

782 The posterior superior iliac spine is a most useful landmark for marrow punctures because, first, the patient cannot easily watch the procedure and, second, a greater volume of marrow can be aspirated from this region than from other red-marrow-containing areas. The skin dimple above the buttock is the surface marking and the needle is inserted 1 cm below and laterally, i.e. in the plane of the hip bone.

783 On viewing this X-ray, an outline resembling a small Scottish terrier can be seen. Its nose is the transverse process, the eye the pedicle, whist the superior articular process forms the tip of the ear. The dog's neck is the 'pars interarticularis' – a radiological term equivalent anatomically to the anterior part of the lamina. Defects of ossification in the pars may lead to spondylolisthesis, a slippage of one vertebra forwards on another.

784 Cervical vertebrae are prone to dislocation due to the almost horizontal alignment of the articular facets.

785 Death. This sudden event may be caused by the fractured dens being driven into the spinal cord. This may be the cause of death when judicial hanging is carried out.

786 To find entry into the vertebral artery above the level of C6 requires more luck than judgement, as the artery then lies in the bony vertebrarterial

foramina until it enters the skull through the foramen magnum. Once inside the skull, it joins its fellow to become the basilar artery lying on the clivus. The vein, however, often stays inside the foramen of C7 unaccompanied by the artery.

787 A posterior prolapse of the C2/3 disc may cause pressure on the spinal cord above the phrenic nerve outflow of C3, 4 and 5. If this happens suddenly respiration will cease and, if not maintained artificially, the patient will die.

788 To perform a cisternal puncture, start with the needle lying just under the occiput and the patient's neck flexed, and advance it in a plane which passes through the external auditory meatus and the nasion. To enter the cisterna magna, the needle must pass through the skin and superficial fascia and through the ligamentum nuchae which is the cranial extension of both the supraspinous and interspinous ligaments. Having pierced the posterior atlanto-occipital membrane, the continuation of the ligamenta flava, the tip lies amongst the posterior vertebral venous plexus of the epidural space. This lies at about 3 cm depth, and to enter the cisterna the needle must be advanced gently through the dura and arachnoid. It is worth remembering that the vertebral arteries lie only 1.5 cm lateral to the midline and that the medulla is just 1 cm or so in advance of the needle tip! Paediatric deaths or injury can happen during this procedure.

789 A prolapsed intervertebral disc is most commonly seen in the lower lumbar region at L4/5 or L5/S1. The consequences include low back pain and shooting pains down the buttocks, posterior thighs and legs towards the foot, often termed sciatica. On examination, the patient often has great difficulty in performing the straight-leg raise test due to pain caused by the test stretching often the sciatic nerve, the path by impingement of the offending disc via nerve root.

B Information answers

790 The erector spinae is better termed the sacrospinalis. It arises from the median sacral crest and ascends in the grooves on either side between the

vertebral spines and their transverse processes. It splits in the lumbar region into three: a lateral group of muscles (iliocostalis), an intermediate group (longissimus) and a medial group (spinalis).

791 The transversospinalis group of muscles is composed of three groups. From superficial to deep, these are the semispinalis group (thoracis, cervicis, capitis), multifidus and the rotatores group (cervicis, thoracis and lumborum). These muscles run upwards and medially from the transverse processes of vertebrae to the vertebral spines. Those of semispinalis span about five interspaces, those of multifidus about three and rotatores one or two.

792 All the deep muscles of the back are supplied by the posterior primary rami of the spinal nerves.

793 The dorsal ramus of C1 (suboccipital nerve) does not have any cutaneous territory, unlike other dorsal rami (but it is sensory to the dura around the sigmoid sinus). The suboccipital nerve does, however, contain motor fibres and supplies the four muscles of the suboccipital triangle (superior oblique, inferior oblique, rectus capitis posterior major and rectus capitis posterior minor).

794 The ventral rami of C2 and 3 contribute to the sensory supply of the back of the neck via the greater auricular nerve (C2, 3) and the lesser occipital nerve (C2). The greater auricular nerve supplies skin over the parotid, mastoid process and posteroinferior aspect of the auricle. The lesser occipital nerve supplies skin over the upper posterior part of sternocleidomastoid. The remainder of the back of the neck is supplied by the posterior rami of the spinal nerves.

795 The small muscles of the suboccipital region are rectus capitis posterior major, rectus capitis posterior minor, inferior oblique and superior oblique. Their main action is to help stabilise the skull and cancel out the unwanted movements of the prime movers (sternocleidomastoid, splenius, capitis, etc.). They are all supplied by the posterior ramus of C1. Flexion and extension movements in this region occur at the atlanto-occipital joint (nodding); rotation movements occur at the atlantoaxial joint (shaking). Rectus capitis posterior

major runs from the occiput to C2 and thus acts over both joints. It is the only muscle to do this, and it extends the head and rotates it to its own side. Rectus capitis posterior minor runs from the occiput to the posterior arch of the atlas. It extends the head. The inferior oblique runs from the spine of the axis to the lateral mass of the atlas. It rotates the skull and atlas to its downside. The superior oblique muscle runs from occiput to the lateral mass of the atlas. It is a lateral flexor of the skull on the atlas.

796 The most vital structure in the suboccipital triangle is the vertebral artery (third part). It exits the foramen transversarium of the atlas and curves backwards behind the lateral mass of the atlas to lie in a groove on the upper surface of the posterior arch of the atlas, and enters the foramen magnum. It is destined to supply the hindbrain – hence its importance.

797 The suboccipital triangle is bounded by the rectus capitis posterior major muscle, the superior oblique muscles and the inferior oblique muscle. Its roof is skin and dense fibrofatty tissue. Below this, its floor is the lateral aspect of the posterior arch of the atlas and the posterior atlanto-occipital membrane.

B Applied answers

798 The auscultation triangle overlies the sixth and seventh intercostal spaces of the back – its boundaries are the vertebral border of the scapula, the lateral border of trapezius, and the upper horizontal border of latissimus dorsi. It is so named because deep to it is the cardiac orifice of the stomach, where the splash of swallowed liquids were timed in cases of oesophageal obstruction in the pre-radiological days! (Its lack of covering muscle also make it a good site to hear chest sounds.)

The lumbar triangle of Petit, the floor of which is the internal oblique muscle, is bounded by latissimus dorsi, external oblique's free posterior border and the iliac crest. Rarely, it is the site of a lumbar hernia.

799 The valveless vertebral venous plexus runs the entire length of the vertebral column and has both

external and internal components which anastomose freely with each other as well as with lumbar, intercostal and pelvic veins. The internal plexus lies within the extradural space and is arranged in four longitudinal channels. The prostatic venous plexus is thought to communicate directly with this venous system, and thus explains the occurrence of early prostatic metastases in the vertebrae and skull.

800 Compression of the sacral plexus will cause sciatica and low back ache. If the dorsal rami are also affected then the erector spinae of the lumbosacral region may go into spasm.

801 The lateral part of the upper arm is supplied by the C5 dermatome. X-rays of the middle and lower cervical vertebrae, especially the C4/5 intervertebral space, might well produce a diagnosis.

802 This is the classic presentation of bilateral cervical ribs causing pressure on the T1 nerve roots. Sometimes these ribs can be seen to articulate with the first rib or may even simulate a lesion in the apex of the lung.

C Information answers

803 The pia mater closely invests the spinal cord. On each side thickenings are formed in it between the nerve roots – the denticulate ligaments. They pass laterally across the subarachnoid space to fuse with the arachnoid and dura mater. They assist the spinal cord's suspension in the centre of the meningeal sheath.

804 The pia mater, having covered the spinal cord throughout its length, continues as a thread from the apex of its termination, through the subarachnoid space to be attached to the coccyx inferiorly. This thread of pia is called the filum terminale.

805 The spinal cord terminates at the level of the lower border of L1 in the adult. The spinal theca terminates at the level of S2 and, therefore, the subarachnoid space ends at this level. Lumbar puncture is carried out between these two levels, usually at L4/5. In the infant, the spinal cord ends at a lower level than in the adult – around L3/4.

806 There are 31 pairs of spinal nerves subdivided as follows: cervical 8, thoracic 12, lumbar 5, sacral 5, coccygeal 1.

807 The spinal cord runs in the vertebral canal surrounded by three membranes. The pia mater closely invests the cord; the dura mater and arachnoid mater are in apposition and lie close to the circumference of the vertebral canal. There is a space between arachnoid and pia – the subarachnoid space. This is filled with shock-absorbing fluid – the cerebrospinal fluid.

808 The dural sheath lies in the vertebral canal but is separated from its walls by the extradural space, which is filled with loose areolar tissue and fat. Embedded in this fascia is a plexus of veins, the internal vertebral venous plexus. The importance of the extradural space is that the spinal nerve roots traverse it and these may be blocked by an injection of local anaesthetic in this region. Furthermore, the venous plexus may be damaged during a traumatic lumbar puncture with a resultant bloody tap.

809 The dura mater sends tubular lateral prolongations which surround each root of the spinal nerves. These prolongations eventually become continuous with the epineurium of the mixed spinal nerve as it passes through the intervertebral foramen. Although short in the upper region of the cord, the prolongations become longer inferiorly because of the increasing obliquity of nerve roots due to difference in growth between cord and spine.

C Applied answers

810 The contrast medium is within the subarachnoid space which extends along the spinal nerve roots, and consequently they are seen on normal X-rays as small lateral extensions.

811 During development, the vertebral column outgrows the spinal cord. In most adults the cord ends at the lower border of L1; that of the newborn infant extends slightly lower, to the third lumbar vertebra. In most infants, and even a few adults, a lumbar puncture performed at L2 may cause laceration of the spinal cord, with subsequent paralysis and sensory loss.

812 In a lateral view, the anterior margin of the thecal sac has slight concavities due to the bulging of the annulus fibrosus of the normal intervertebral discs. These are normally checked – especially those of the lumbar region – for herniation, in patients with back pain and sciatica.

813 Particularly in the cervical region, the intervertebral foramen is relatively narrow to accommodate its spinal vessels and nerve. As the foramen is not capable of expansion due to its bony walls, compression – particularly of the nerve – can easily happen. This gives rise to root pains felt over the appropriate dermatomes.

814 With the patient sitting or lying in the fetal position, a lumbar puncture needle entering the back will pierce, in succession: skin, subcutaneous tissues, supraspinous and interspinous ligaments, and the ligamentum flavum if the needle is not absolutely midline. The elastic ligamentum flavum is felt as a definite resistance. The needle will then cross the epidural space containing the posterior intervertebral venous plexus and thence through the dura and arachnoid mater (spinal theca) to enter the subarachnoid space. Here the needle tip will be bathed in CSF and lie amongst the lumbosacral nerve roots of the cauda equina.

Biographical notes on eponyms

ADDISON, Christopher (Viscount Addison of Stallingborough) 1869–1951
Professor in Anatomy, Sheffield, then Dean of Charing Cross Hospital, then Lecturer at St Bartholomew's Hospital (Barts medical student). In 1910 elected MP for the Hoxton division of Shoreditch. He was the first Minister of Health (1918–21). *Transpyloric Plane of Addison*.

ALCOCK, Benjamin 1801–?
In 1849 was appointed Professor of Anatomy in Queen's College, Cork, but was called upon to resign in 1853 due to a dispute concerning the working of the Anatomy Acts. Went to USA in 1855, and was heard of no more! *Alcock's canal* – for the internal pudendal vessels and pudendal nerve in the ischioanal fossa.

BAKER, William M. 1839–1896
An English surgeon. *Baker's cyst* – a cyst of the bursa of the popliteal fossa.

BARTHOLIN, Caspar (Secundus) 1655–1738
Succeeded his father, Thomas Bartholin, as Professor of Medicine, Anatomy and Physics in Copenhagen. *Bartholin's ducts* – sublingual ducts that open into the submandibular duct. *Bartholin's glands* – greater vestibular glands.

BATSON, Oscar Vivian 1894–1979
Professor of Anatomy, University of Pennsylvania. Vertebral venous plexus of *Batson*. A valveless plexus implicated in bony metastatic spread of carcinomata – particularly of the prostate.

BELL, Sir Charles 1774–1842
Surgeon, anatomist and artist. Founder of the Middlesex Hospital Medical School. Later Professor of Surgery, Edinburgh. *Long thoracic nerve of Bell* to serratus anterior. *Bell's palsy* – facial nerve paralysis of unknown aetiology.

BIGELOW, Henry Jacob 1818–1890
Professor of Surgery at Harvard, USA. *The iliofemoral ligament of Bigelow* (hip joint).

BOCHDALEK, Vincent 1801–1883
Anatomist at Prague. *Bochdalek's foramen* – pleuroperitoneal canal. *Bochdalek's gap* – lumbocostal trigone. *Bochdalek's gland* – derived from primitive thyroglossal duct.

BOYDEN, Edward 1886–1978
American anatomist. *Boyden's sphincter* – the separate sphincter of the common bile duct proximal to the sphincter of Oddi.

BUCK, Gurdon 1807–1877
New York surgeon. *Buck's fascia* – deep penile fascia.

CALOT, Jean François 1861–1944
French surgeon. *Calot's triangle* – bounded by the common hepatic and cystic ducts and the liver.

CAMPER, Pieter 1722–1789
Professor of Medicine, Anatomy, Surgery and Botany in Groningen, Holland. *Camper's fascia* – fatty layer of the superficial fascia of the abdomen.

CHIPPENDALE 1968–.
Male body-building stage performer.

COLLES, Abraham 1773–1843
Professor of Anatomy and Surgery in Dublin. *Colles' fascia* – perineal fascia and deep layer of the superficial fascia of the abdomen. *Colles' fracture* – of the extremity of the radius. *Colles' ligament* reflected inguinal ligament.

COOPER, Sir Astley 1768–1841
English surgeon, Guy's Hospital. *Cooper's fascia* – cremasteric fascia. *Suspensory ligaments of Cooper* – fibrous tissue in the breast which accounts for dimpling of the skin in certain carcinomata.

CORTI, Alfonso 1822–1888
Italian anatomist. *Organ of Corti* – Specialised cells within the cochlear duct.

COWPER, William 1666–1709
London surgeon. FRS 1698. *Cowper's glands* – bulbourethral glands.

DOUGLAS, James 1675–1742
Scottish anatomist and 'man-midwife' who lived in London. Physician to the Queen. FRS 1706. *Arcuate line of Douglas* – of the posterior layer of the rectus sheath. *Pouch of Douglas* – rectouterine peritoneal pouch.

DENONVILLIERS, Charles Pierre 1808–1872
Professor of Anatomy and Surgery in Paris. *Denonvillier's fascia* – rectovesical fascia in the male.

DUPUYTREN, Guillaume 1777–1835
French surgeon and pathologist. *Dupuytren's contracture* – fibroses of palmar fascia causing fixed flexion deformity usually of ring and little fingers.

EDINGER, Ludwig 1855–1918
German anatomist. *Edinger–Westphal nucleus* – midbrain group of cells forming part of the oculomotor nerve from which originate the preganglionic parasympathetic fibres to the eye.

ERB, Wilhelm 1840–1921
German neurologist. *Erb's palsy* – damage to the upper trunk of the brachial plexus.

EUSTACHIO (EUSTACHI, EUSTACHIUS), Bartolomeo 1513–1574 (DOB uncertain)
Professor of Anatomy in Rome and Physician to the Pope. *Eustachian tube* – the auditory tube (strictly, its cartilaginous part). *Eustachian valve* – of the inferior vena cava.

FALLOPIUS (FALLOPIO), Gabriele 1523–1563
Professor of Anatomy and Surgery in Padua. *Fallopius' canal* – facial nerve canal. *Fallopian tube* – uterine tube.

FRIEDREICH, Nikolas 1825–1882
German neurologist. *Friedreich's ataxia* – hereditary spinal ataxia.

FROMENT, Jules 1878–1946
Professor of Medicine, Lyons, France. *Froment's sign* – seen in ulnar nerve palsy when the distal phalanx of thumb flexes whilst trying to hold a piece of paper against the index finger. This is due to adductor pollicis weakness permitting overaction of flexor pollicis longus.

GALEN, Claudius c. 130–200
Second century physician and gatherer of medical information. Many of his erroneous ideas were regarded as gospel until the fifth and sixth centuries. *Great vein of Galen* – internal cerebral vein.

GARTNER, Hermann Treschow 1785–1827
Copenhagen anatomist. *Gartner's canal or duct* – the vestige of the mesonephric duct found in the broad ligament of the uterus and sometimes extending in the wall of the vagina to the vulva.

GUYON, Felix Jean Casimir 1831–1920
Genitourinary surgeon and professor of surgical pathology in Paris. *Guyon's canal* – for the ulnar nerve beside the pisiform bone.

HESSELBACH, Franz K. 1759–1816
German surgeon. *Hesselbach's fascia* – cribriform fascia. *Hesselbach's triangle* – inguinal triangle.

HIGHMORE, Nathaniel 1613–1685
English anatomist. *Highmore's antrum* – of the maxilla.

HILTON, John 1804–1878
Surgeon at Guy's Hospital, London, from 1849–1871. *Hilton's law* – a nerve which supplies a muscle that crosses a joint, gives a branch to it and to the skin over that joint.

HORNER, Johann F. 1831–1886
A Zürich ophthalmologist. *Horner's syndrome* – facial and ophthalmic paresis, due to cervical sympathetic interruption.

HOUSTON, John 1802–1845
Lecturer in Surgery in Dublin, and Physician to the City Hospital. *Valves of Houston* – transverse rectal folds.

HUNT, James RAMSAY 1874–1937
American neurologist. *Ramsay Hunt's syndrome* – facial paralysis, otalgia, aural herpes due to geniculate ganglion herpes zoster.

HUNTER, John 1728–1793
Scottish anatomist and surgeon who worked in London. Founder of the Hunterian Museum, now in the custody of the Royal College of Surgeons of England. *Hunter's canal* – adductor canal or subsartorial canal of the thigh.

KIESSELBACH, Wilhelm 1839–1902
German laryngologist. *Kiesselbach's area* – anterior portion of nasal septum often site of epistaxis (syn: Little's area).

KILLIAN, Gustav 1860–1921
Berlin Professor of Otolaryngology. *Killian's bundle* – lowest fibres of inferior constrictor of pharynx. *Killian's dehiscence* – weakness between fibres of inferior pharyngeal constrictor.

KLUMPKE, Augusta Dejerine 1859–1927
French neurologist. *Klumpke's paralysis* – paralysis of the intrinsic muscles of the hand, often due to birth injury.

LATARJET, André 1877–1947
French anatomist. *Latarjet's nerve* – a branch of the anterior vagus to the upper border of the pylorus and antrum of the stomach.

LITTLE, James Laurence 1836–1885
Professor of Surgery, University of Vermont. *Little's area* – site of haemorrhage on the nasal septum.

LOUIS, Antoine 1723–1792
Surgeon and physiologist of Paris whom American authorities cite as noting the 'angle' but this is not mentioned in this works. *Angle of Louis* – the manubriosternal junction.

MACKENRODT, Alwin 1859–1925
Professor of Gynaecology in Berlin. *Mackenrodt's ligament* – cardinal (lateral cervical) ligament of the uterus.

McBURNEY, Charles 1845–1914
New York surgeon. *McBurney's point* – between 1½ and 2 inches (3.8–5.0 cm) from the right anterior superior iliac spine upon a line to the umbilicus. The given site of maximum tenderness with appendicitis.

MECKEL, Johann Friedrich 1714–1774
Professor of Anatomy, Botany and Gynaecology in Berlin. *Meckel's space/cave* – the trigeminal cavum. The subarachnoid space around cranial nerve V as it lies in the middle cranial fossa.

MECKEL, Johann Friedrich the younger 1781–1833
Grandson of the preceeding. A German comparative anatomist and embryologist. *Meckel's cartilage* – the

cartilage of the first branchial arch. *Meckel's diverticulum* – a diverticulum of the ileum; a persistent proximal part of the vitellointestinal duct.

MEIBOMIUS (MEIBOM) Hendrik 1638–1700
Professor of Medicine at Helmstadt and later Professor of Poetic Art. *Meibomian cyst* chalazion due to infection of a tarsal gland. *Meibomius' gland* – tarsal gland.

MONTGOMERY, William 1797–1859
Professor of Midwifery in Dublin. *Montgomery's glands or tubercles* – elevated reddish areolar glands easily seen in pregnancy.

MORGAGNI, Giovanni Battista 1682–1771
A Padua anatomist and pathologist. *Appendix or hydatid of Morgagni* – appendix testis or vesicular appendix of epoöphoron. *Columns of Morgagni* – anal columns. *Foramen of Morgagni* – foramen caecum of the tongue. *Lacunae/Fossa of Morgagni* – fossa navicularis.

MÜLLER, Heinrich 1820–1864
German anatomist. *Müller's muscle* – superior tarsal muscle of the eyelids, or circular fibres of the iris.

MÜLLER, Johannes Peter 1801–1858
Professor of Anatomy and Physiology in Berlin. *Müller's duct* – primordial female genital duct – the paramesonephric duct.

NUCK, Anton 1650–1692
Professor of Anatomy and Medicine in Leyden. *Canal of Nuck* – patent processus vaginalis peritonei in the female.

ODDI, Ruggero 1864–1913
An Italian physician. *Sphincter of Oddi* – sphincteric muscle fibres around the termination of the bile duct and main pancreatic duct.

PAGET, Sir James 1814–1899
English surgeon. *Paget's disease of nipple* – malignant change in the skin surrounding the nipple.

PANCOAST, Henry Khunrath 1875–1939
Professor of Radiology in Pennsylvania, USA. *Pancoast's syndrome* – apical carcinoma of lung associated with Horner's syndrome, brachial plexus palsy, pain down the arm and distension of the neck veins.

PETIT, Jean Louis 1674–1750
Paris surgeon. *Petit's triangle* – lumbar triangle.

PEYER, Johan Konrad 1653–1712
Professor of Logic, Rhetoric and Medicine in Schaffhausen, Switzerland. *Peyer's patches* – aggregated lymphoid follicles in the lower ileum.

PFANNENSTIEL, Hermann Johann 1862–1909
Pfannenstiel incision – transverse suprapubic incision used for pelvic surgery.

POTT, Percival 1713–1788
English surgeon. *Pott's disease* – tuberculosis of the spine. *Pott's fracture* – ankle fractures involving the malleoli and ligaments.

POUPART, François 1661–1709
Surgeon to the Hôtel Dieu, Paris. *Poupart's ligament* – inguinal ligament.

RAYNAUD, Maurice 1834–1881
Parisian physician. *Raynaud's disease* – peripheral vascular disturbance consisting of spasmodic contractions of the digital arteries.

REISSNER, Ernst 1824–1878
German anatomist. *Reissner's membrane* – vestibular membrane of the inner ear, dividing scala vestibuli and scala media.

RIEDEL, Berhard 1846–1916
German surgeon. *Riedel's lobe* – a tongue-like extension of the right lobe of the liver, thought at one time to be due to the wearing of tight corsets – now known to be a developmental anomaly!

SANTORINI, Giovanni 1681–1737
Italian anatomist. *Cartilage of Santorini* – corniculate cartilage. *Duct of Santorini* – accessory pancreatic duct.

SCARPA, Antonio 1747–1832
Venetian anatomist, orthopaedist and ophthalmologist. *Scarpa's fascia* – deep layer of the superficial fascia of the lower abdomen. *Scarpa's nerve* – nasopalatine nerve. *Scarpa's triangle* – femoral triangle.

SCHLEMM, Friedrich 1795–1858
German anatomist. *Canal of Schlemm* – venous sinus of the sclera found at the corneoscleral junction.

SIBSON, Francis 1814–1876
English anatomist. *Sibson's fascia* – the suprapleural membrane.

SKENE, Alexander Johnston Chalmers 1838–1900
Born in Aberdeen, qualified in Michigan. Professor of Gynaecology in the Long Island College Hospital, Brooklyn, New York. *Skene's glands* – the paraurethral glands in the female.

SPIGELIUS, Adrian (van der Speighel) 1578–1625
Belgian anatomist in Padua. *Spigelian hernia* – herniation along the linea semilunaris of the abdomen. *Spigelius's lobe* – caudate lobe of liver.

STENSON, Niels or Nicholaus 1638–1686
Danish anatomist, who later became a minister and a bishop. *Stenson's duct* – duct of the parotid gland lying across masseter muscle.

TENON, Jacques René 1724–1816
Professor of Pathology in the Academy of Sciences, Paris and Chief Surgeon at the Salpetrier. *Tenon's capsule or fascia* – the fascial sheath of the eyeball.

THEBESIUS, Adam 1686–1732
German physician. *Thebesius' valve* – valve of the coronary sinus. *Thebesius' veins* – small veins draining directly into the cardiac chambers.

TROISIER, C. E. 1844–1919
French Pathology Professor in Paris. *Troisier's sign or node* – enlargement of left supraclavicular node due to malignancy in the stomach or lung.

TRENDELENBURG, Freidrich 1844–1924
German surgeon. *Trendelenburg's position* – operating position with head down and pelvis elevated, as used in the treatment of shock. *Trendelenburg's sign* – dipping of the pelvis in congenital hip dislocation.

VATER, Abraham 1684–1751
Professor of Anatomy, Botany, Pathology and Therapeutics in Wittenberg, Germany. *Ampulla of Vater* – hepaticopancreatic ampulla.

VIDIUS (Guido Guidi) 1500–1569
Italian anatomist and physician to Francis I of France; from 1548, Professor of Medicine at Pisa. *Vidian canal* – pterygoid canal. *Vidian nerve* – nerve of pterygoid canal.

VIRCHOW, Rudolf Ludwig Karl 1821–1902
Professor of Pathological Anatomy, University of Wurzburg, and later in Berlin. *Virchow's nodes* – enlarged left supraclavicular lymph nodes indicating intraperitoneal malignancy.

VOLKMANN, Richard von 1830–1889
German Professor of Surgery. *Volkmann's contracture* – tissue loss, especially muscular, due to ischaemia. *Volkmann's splint* – guttered splint for lower extremity fractures.

WALDEYER, Heinrich von 1836–1921
Professor of Pathological Anatomy in Breslau, and later in Berlin. *Waldeyer's ring* – lymphatic ring in the pharynx. *Waldeyer's fascia* – posterior rectal fascial septum.

WESTPHAL, Karl 1833–1890
German neurologist. *Edinger-Westphal nucleus* – group of cells in midbrain forming part of the oculomotor nerve from which preganglionic parasympathetic fibres to the eye are derived.

WHARTON, Thomas 1614–1673
English anatomist. *Wharton's duct* – main submandibular duct opening at the frenulum of the tongue. *Wharton's jelly* – connective tissue (mucous) of umbilical cord.

WINSLOW, Jacob Benignus 1669–1760
Danish anatomist who, at the age of 74, was appointed Professor of Anatomy, Physics and Surgery in Paris. *Foramen of Winslow* – omental foramen. *Ligament of Winslow* – oblique popliteal ligament.

WIRSUNG, Johann 1600–1643
German anatomist in Padua. *Wirsung's duct* – the main pancreatic duct.

WOLFF, Kaspar Friedrich 1733–1794
Professor of Anatomy and Physiology in St Petersburg. One of the founders of modern embryology. *Wolffian duct* – mesonephric duct. *Wolffian body* – mesonephros.